Richard DesJardins
4/7/04

Lens Design

OPTICAL ENGINEERING

Founding Editor

Brian J. Thompson

Distinguished University Professor
Professor of Optics
Provost Emeritus
University of Rochester
Rochester, New York

Editorial Board

Toshimitsu Asakura
Hokkai-Gakuen University
Sapporo, Hokkaido, Japan

Nicholas F. Borrelli
Corning, Inc.
Corning, New York

Chris Dainty
Imperial College of Science,
Technology, and Medicine
London, England

Bahram Javidi
University of Connecticut
Storrs, Connecticut

Mark Kuzyk
Washington State University
Pullman, Washington

Hiroshi Murata
The Furukawa Electric Co., Ltd.
Yokohama, Japan

Edmond J. Murphy
JDS/Uniphase
Bloomfield, Connecticut

Dennis R. Pape
Photonic Systems Inc.
Melbourne, Florida

Joseph Shamir
Technion–Israel Institute
of Technology
Hafai, Israel

David S. Weiss
Heidelberg Digital L.L.C.
Rochester, New York

1. Electron and Ion Microscopy and Microanalysis: Principles and Applications, *Lawrence E. Murr*
2. Acousto-Optic Signal Processing: Theory and Implementation, *edited by Norman J. Berg and John N. Lee*
3. Electro-Optic and Acousto-Optic Scanning and Deflection, *Milton Gottlieb, Clive L. M. Ireland, and John Martin Ley*
4. Single-Mode Fiber Optics: Principles and Applications, *Luc B. Jeunhomme*
5. Pulse Code Formats for Fiber Optical Data Communication: Basic Principles and Applications, *David J. Morris*
6. Optical Materials: An Introduction to Selection and Application, *Solomon Musikant*
7. Infrared Methods for Gaseous Measurements: Theory and Practice, *edited by Joda Wormhoudt*
8. Laser Beam Scanning: Opto-Mechanical Devices, Systems, and Data Storage Optics, *edited by Gerald F. Marshall*
9. Opto-Mechanical Systems Design, *Paul R. Yoder, Jr.*
10. Optical Fiber Splices and Connectors: Theory and Methods, *Calvin M. Miller with Stephen C. Mettler and Ian A. White*
11. Laser Spectroscopy and Its Applications, *edited by Leon J. Radziemski, Richard W. Solarz, and Jeffrey A. Paisner*
12. Infrared Optoelectronics: Devices and Applications, *William Nunley and J. Scott Bechtel*
13. Integrated Optical Circuits and Components: Design and Applications, *edited by Lynn D. Hutcheson*
14. Handbook of Molecular Lasers, *edited by Peter K. Cheo*
15. Handbook of Optical Fibers and Cables, *Hiroshi Murata*
16. Acousto-Optics, *Adrian Korpel*
17. Procedures in Applied Optics, *John Strong*
18. Handbook of Solid-State Lasers, *edited by Peter K. Cheo*
19. Optical Computing: Digital and Symbolic, *edited by Raymond Arrathoon*
20. Laser Applications in Physical Chemistry, *edited by D. K. Evans*
21. Laser-Induced Plasmas and Applications, *edited by Leon J. Radziemski and David A. Cremers*
22. Infrared Technology Fundamentals, *Irving J. Spiro and Monroe Schlessinger*
23. Single-Mode Fiber Optics: Principles and Applications, Second Edition, Revised and Expanded, *Luc B. Jeunhomme*
24. Image Analysis Applications, *edited by Rangachar Kasturi and Mohan M. Trivedi*
25. Photoconductivity: Art, Science, and Technology, *N. V. Joshi*
26. Principles of Optical Circuit Engineering, *Mark A. Mentzer*
27. Lens Design, *Milton Laikin*
28. Optical Components, Systems, and Measurement Techniques, *Rajpal S. Sirohi and M. P. Kothiyal*
29. Electron and Ion Microscopy and Microanalysis: Principles and Applications, Second Edition, Revised and Expanded, *Lawrence E. Murr*

30. Handbook of Infrared Optical Materials, *edited by Paul Klocek*
31. Optical Scanning, *edited by Gerald F. Marshall*
32. Polymers for Lightwave and Integrated Optics: Technology and Applications, *edited by Lawrence A. Hornak*
33. Electro-Optical Displays, *edited by Mohammad A. Karim*
34. Mathematical Morphology in Image Processing, *edited by Edward R. Dougherty*
35. Opto-Mechanical Systems Design: Second Edition, Revised and Expanded, *Paul R. Yoder, Jr.*
36. Polarized Light: Fundamentals and Applications, *Edward Collett*
37. Rare Earth Doped Fiber Lasers and Amplifiers, *edited by Michel J. F. Digonnet*
38. Speckle Metrology, *edited by Rajpal S. Sirohi*
39. Organic Photoreceptors for Imaging Systems, *Paul M. Borsenberger and David S. Weiss*
40. Photonic Switching and Interconnects, *edited by Abdellatif Marrakchi*
41. Design and Fabrication of Acousto-Optic Devices, *edited by Akis P. Goutzoulis and Dennis R. Pape*
42. Digital Image Processing Methods, *edited by Edward R. Dougherty*
43. Visual Science and Engineering: Models and Applications, *edited by D. H. Kelly*
44. Handbook of Lens Design, *Daniel Malacara and Zacarias Malacara*
45. Photonic Devices and Systems, *edited by Robert G. Hunsperger*
46. Infrared Technology Fundamentals: Second Edition, Revised and Expanded, *edited by Monroe Schlessinger*
47. Spatial Light Modulator Technology: Materials, Devices, and Applications, *edited by Uzi Efron*
48. Lens Design: Second Edition, Revised and Expanded, *Milton Laikin*
49. Thin Films for Optical Systems, *edited by François R. Flory*
50. Tunable Laser Applications, *edited by F. J. Duarte*
51. Acousto-Optic Signal Processing: Theory and Implementation, Second Edition, *edited by Norman J. Berg and John M. Pellegrino*
52. Handbook of Nonlinear Optics, *Richard L. Sutherland*
53. Handbook of Optical Fibers and Cables: Second Edition, *Hiroshi Murata*
54. Optical Storage and Retrieval: Memory, Neural Networks, and Fractals, *edited by Francis T. S. Yu and Suganda Jutamulia*
55. Devices for Optoelectronics, *Wallace B. Leigh*
56. Practical Design and Production of Optical Thin Films, *Ronald R. Willey*
57. Acousto-Optics: Second Edition, *Adrian Korpel*
58. Diffraction Gratings and Applications, *Erwin G. Loewen and Evgeny Popov*
59. Organic Photoreceptors for Xerography, *Paul M. Borsenberger and David S. Weiss*
60. Characterization Techniques and Tabulations for Organic Nonlinear Optical Materials, *edited by Mark Kuzyk and Carl Dirk*

61. Interferogram Analysis for Optical Testing, *Daniel Malacara, Manuel Servín, and Zacarias Malacara*
62. Computational Modeling of Vision: The Role of Combination, *William R. Uttal, Ramakrishna Kakarala, Sriram Dayanand, Thomas Shepherd, Jagadeesh Kalki, Charles F. Lunskis, Jr., and Ning Liu*
63. Microoptics Technology: Fabrication and Applications of Lens Arrays and Devices, *Nicholas F. Borrelli*
64. Visual Information Representation, Communication, and Image Processing, *Chang Wen Chen and Ya-Qin Zhang*
65. Optical Methods of Measurement: Wholefield Techniques, *Rajpal S. Sirohi and Fook Siong Chau*
66. Integrated Optical Circuits and Components: Design and Applications, *edited by Edmond J. Murphy*
67. Adaptive Optics Engineering Handbook, *edited by Robert K. Tyson*
68. Entropy and Information Optics, *Francis T. S. Yu*
69. Computational Methods for Electromagnetic and Optical Systems, *John M. Jarem and Partha P. Banerjee*
70. Laser Beam Shaping: Theory and Techniques, *edited by Fred M. Dickey and Scott C. Holswade*
71. Rare Earth Doped Fiber Lasers and Amplifiers: Second Edition, Revised and Expanded, *edited by Michel J. F. Digonnet*
72. Lens Design: Third Edition, Revised and Expanded, *Milton Laikin*

Additional Volumes in Preparation

Handbook of Optical Engineering, *edited by Daniel Malacara and Brian J. Thompson*

Lens Design
Third Edition, Revised and Expanded

Milton Laikin
Laikin Optical Corporation
Marina del Rey, California

MARCEL DEKKER, INC.　　　　NEW YORK • BASEL

Marcel Dekker, Inc., and the author make no warranty with regard to the accompanying software, its accuracy, or its suitability for any purpose other than as described in the preface. This software is licensed solely on an "as is" basis. The only warranty made with respect to the accompanying software is that the diskette medium on which the software is recorded is free of defects. Marcel Dekker, Inc., will replace a diskette found to be defective if such defect is not attributable to misuse by the purchaser or his agent. The defective diskette must be returned within ten (10) days to:

Customer Service
Marcel Dekker, Inc.
P.O. Box 5005
Cimarron Road
Monticello, NY 12701
(914) 796-1919

ISBN: 0-8247-0507-6

This book is printed on acid-free paper.

Headquarters
Marcel Dekker, Inc.
270 Madison Avenue, New York, NY 10016
tel: 212-696-9000; fax: 212-685-4540

Eastern Hemisphere Distribution
Marcel Dekker AG
Hutgasse 4, Postfach 812, CH-4001 Basel, Switzerland
tel: 1-61-261-8482; fax: 41-1-261-8896

World Wide Web
http://www.dekker.com

The publisher offers discounts on this book when ordered in bulk quantities. For more information, write to Special Sales/Professional Marketing at the headquarters address above.

Copyright © 2001 by Marcel Dekker, Inc. All Rights Reserved.
Neither this book nor any part may be reproduced or transmitted in any form or by any means, electronic or mechanical, including photocopying, microfilming, and recording, or by any information storage and retrieval system, without permission in writing from the publisher.

Current printing (last digit):
10 9 8 7 6 5 4 3 2

PRINTED IN THE UNITED STATES OF AMERICA

To my wife Pat

From the Series Editor

In a recent editorial in the journal *Optical Engineering*, the editor, Roger Lessard, asked, "Are we really afraid of basic optics?" [1]. By *basic optics* he was referring to lens design, optical system design, and instrumental optics. Dr. Lessard explained his concern, which is shared by many educators and employers, that these topics in basic optics are not properly included in the various curricula that purport to provide a rounded education in optical science and engineering. The not unrelated concern is that students do not particularly wish to study these areas, preferring to concentrate on areas of optical science and engineering that are considered "hot." The net result is that there is a shortage of graduates who are trained and educated in lens design and optical system design.

The preceding discussion highlights the need for texts that can be used by practicing engineers to "get up to speed" in the science and technology of lens design. Milton Laikin has provided such a text in an innovative format; this book first appeared in 1991 and a second edition was published in 1995. We are now pleased to present the third edition of this volume, revised and expanded to include a number of new designs and revision of previous designs to include replacement glasses.

The author's preface should be read carefully since it provides the basic philosophy of the book and also details the changes in this edition.

Brian J. Thompson

REFERENCE

1. Lessard, R. A. (2000). *Optical Engineering, 39* 11, p. 2845.

Preface

Of the several very fine texts on optical engineering, none gives detailed design information or design procedures for a wide variety of lens systems. This text is written as an aid for the practicing lens designer as well as for those aspiring to be lens designers.

It is assumed tht the reader is familiar with ray tracing procedures, paraxial data, and third-order aberrations, and of course has available a computer lens design and analysis program. (See Apprendix D for a list of commercially available lens design programs.) As the personal computer has increased in popularity and computing power, it exceeds, in its scientific computing ability, the large computers of the 1960s through '80s. Many excellent programs are now available for lens optimization, ray trace analysis, lens plotting, modulation transfer function (MTF) computations, and other applications. All these programs, however, are optimization programs; the designer must input a starting solution. The most common lens design question asked in my courses in geometric optics at the University of California at Los Angeles is "How did you arrive at the starting design?" One of the purposes of this text is to answer just that question.

All optical glasses listed in the designs are from the Schott glass catalog except for the Ohara FPL-53 element used in the designs shown in Figs. 2-3, 7-3, and 7-5. Other glass manufactuers (Ohara, Hoya, Chance, Corning, etc.) make nearly equivalent types of glass. This was done for convenience; I do not endorse any one glass manufacturer. In some of the prescriptions the material listed is quartz. This is fused quartz (not the crystal material) and is referred to in some programs (ZEMAX) as silica.

All lens prescription data are given full size in inches. This allows a practical system for presenting a particular application (perhaps for a 35 mm reflex camera). The lens diameters then have reasonable values of edge thickness. The usual sign convention applies; thickness is an axial dimension to the next surface and radius is + if the center of curvature is to the right of the surface. Light travels left to right, from the long conjugate to the short. Lens diameters are not necessarily clear apertures, but rather the actual lens diameters as shown in the lens diagrams.

All data for the visual region are centered at the e line, and cover F′ to C′. Two infrared regions are considered: 8–14 µm (center at 10.2) and 3.2–4.2 µm (center at 3.63). All data for the ultraviolet are centered at 0.27 µm and cover 0.2–0.4 µm. Field of view (FOV) is quoted in degrees and applies to the full field.

In the first edition of this book, all calculations for the fixed focal length designs were performed using David Grey's optical design and analysis program POP (or COVLY for the cylindrical and decentered systems). The orthonormalization technique is described by Grey (1966) and in the program by Walters (1966). COVLY uses orthogonal polynomials as aberration coefficients (Grey, 1980).

Calculations and plotting for zoom lenses, cylindrical and prism systems, and gradient index lenses were performed using the ZEMAX progam (Moore, 1999). Many of the fixed focal length lenses were also designed and analyzed using ZEMAX. The zoom movement plots were done using the Zoom Spacings program in Appendix E.

An added feature to this edition is the inclusion of a disk containing 126 lens prescriptions, the same ones listed in the text. By using data directly from the computer, some of the prescription errors found in earlier editions have been eliminated. The format is described in Appendix E (radius, thickness, material, diameter). Files correspond to the figures in the text (not table numbers).

On the MTF plots, the angle quoted is semifield in degrees as seen from the first lens surface (or in some cases, object or image heights.) The data are diffraction-included MTF. MTF was selected as a method of analysis because

1. Diffraction effects are included.
2. Image resolution can readily be deduced.
3. MTF data are often required by the systems designer.

This use of MTF has the obvious disadvantage of not providing the same insight into lens aberrations as the traditional ray intercept plots.

Distortion is defined as

$$D = \frac{Y_c - Y_g}{Y_g}$$

where Y_c is the actual image height at full field and Y_g is the corresponding paraxial image height. For afocal systems,

$$D = \frac{\tan\theta' / \tan\theta - m}{m}$$

where θ' is the emerging angle at full field and m is the paraxial magnification (Kingslake, 1965).

Although I have made a great deal of effort to assure accuracy, the user of any of these designs should:

1. Carefully analyze the prescription to be sure that it meets the particular requirements.
2. Check the patent literature for possible infringement. Copies of patents are available ($3.00 each) from the Commissioner of Patents, Washington, D.C. 20231. Patent literature is a valuable source of detail design data. Since patents are valid for a period of 20 years after the application has been received by the Patent Office (previously 17 years after the patent was granted), older patents are now in the public domain and so may be freely used (U.S. Patent Office, 1999).

The Internet may be utilized to perform a search of the patent literature. The patent office site is http:/www.uspto.gov. Patents may be searched by number or subject. Text and lens prescriptions are given (only for patents issued after 1976) but not lens diagrams or other illustrations, which are available only in the printed version of the patent. In *Optics and Photonics News*, Brian Caldwell discusses a patent design each month. There are also patent reviews in *Applied Optics*.

3. All data for lens prescriptions are rounded, so a very slight adjustment in back focal length (BFL) may be necessary (particularly with lenses of low f number) if the reader desires to verify the MTF data.

Although the male gender is used (for simplicity) throughout this book, it is not to be implied that only men are optical engineers. More and more women are being attracted to this field and I acknowledge their contributions.

This edition contains minor corrections to the previous editions and presents several new designs. Some of the glasses used in the previous designs are now considered obsolete and were replaced.

Since most available computer design programs (see Appendix D) now have glass catalogs incorporated into their programs, the refractive index tables of the previous editions have been deleted. The Refractive Index program described in Appendix E may be used to compute the refractive index of glasses from several manufacturers as well as of some infrared materials, plastics, sea water, etc.

Also included is a disk containing, in addition to lens prescriptions, several programs that the reader might find useful in optical design. These are not lens optimization programs but rather short programs to, for example, calculate an achromat, perform third-order triplet design, do lens drawings, and perform refractive index calculations. They are described in Appendix E.

I would like to thank Kenneth Moore, James Sutter, and Mary Turner of Focus Software for their help with the ZEMAX program.

Milton Laikin

REFERENCES

Grey, D. S. (1966). Recent developments in orthonormalization of parameter space, in *Lens Design with Large Computers*, Proceedings of the Conference, Rochester, NY, Institute of Optics, Paper 2.

Grey, D.S. (1980). Orthogonal polynomials as lens aberration coefficients. *Proceedings of the 1980 International Lens Design Conference*, SPIE, p. 85.

Kingslake, R. (1965). *Applied Optics and Optical Engineering*, Vol. 1, Academic Press, New York, p. 239.

Moore, K. (1999). *ZEMAX Optical Design Program, User's Guide*, Focus Software, Inc., Tucson, AZ.

U.S. Patent Office (1999). General Information Concerning Patents, Available from U.S. Govt. Printing Office, Supt. of Documents, Washington, DC 20402.

Walters, R. M. (1966). Odds and ends from a Grey box, in *Lens Design with Large Computers, Proceedings of the Conference*, Rochester, NY, Paper 3.

Contents

From the Series Editor Brian J. Thompson		*v*
Preface		*vii*
List of Lens Designs		*xvii*
1.	The Method of Lens Design	1
2.	The Achromatic Doublet	45
3.	The Air-Spaced Triplet	55
4.	Triplet Modifications	65
5.	Petzval Lenses	71
6.	Double Gauss Lenses	75
7.	Telephoto Lenses	85
8.	Inverted Telephoto Lenses	97
9.	Very Wide Angle Lenses	105
10.	Eyepieces	121
11.	Microscope Objectives	133

12. In-Water Lenses	145
13. Afocal Optical Systems	157
14. Relay Systems	169
15. Catadioptric and Mirror Optical Systems	181
16. Periscope Systems	203
17. IR Lenses	211
18. Ultraviolet Lenses and Optical Lithography	221
19. F Theta Scan Lenses	233
20. Endoscopes	241
21. Enlarging Lenses	245
22. Projection Lenses	249
23. Telecentric Systems	265
24. Laser-Focusing Lenses	271
25. Heads Up Display	279
26. Achromatic Wedges	287
27. Wedge Plates and Rotary Prisms	291
28. Anamorphic Attachments	297
29. Illumination Systems	305
30. Lenses for Aerial Photography	313

31. Radiation-Resistant Lenses	323
32. Lenses for Microprojection	327
33. First-Order Theory, Mechanically Compensated Zoom Lenses	331
34. First-Order Theory, Optically Compensated Zoom Lenses	337
35. Mechanically Compensated Lenses	341
36. Optically Compensated Zoom Lenses	387
37. Copy Lenses with Variable Magnification	401
38. Variable Focal Length Lenses	409
39. Gradient Index Lenses	419

Appendices

A. Film Formats	429
B. Flange Distances	431
C. Thermal and Mechanical Properties	433
D. Commercially Available Lens Design Programs	435
E. Program Optics	437

Index — *471*

List of Lens Designs

2-1 48 inch focal length achromat
2-2 Cemented achromat
2-3 Cemented achromat, 10 inch focal length
2-4 Cemented apochromat
3-2 5 inch focal length f/3.5 triplet
3-3 IR Triplet 8–14 micron
3-4 4 inch IR triplet 3.2–4.2 micron
3-5 50 inch, f/8 triplet
4-1 Heliar f/5
4-2 Tessar lens 4 inch, f/4.5
4-3 Slide projector lens
5-1 Petzval lens f/1.4
5-2 100mm f/2.8 Petzval lens
6-1 Double Gauss f/2.5
6-2 50 mm f/1.8 SLR camera lens
6-3 f/1 5 deg FOV double Gauss
6-4 25 mm f/.85 double Gauss
7-2 Telephoto f/5.6
7-3 f/2.8 180 mm telephoto
7-4 400 mm f/4 telephoto
7-5 1000 mm f/11 telephoto
7-6 Cemented achromat with 2 × extender
8-1 Inverted telephoto f/3.5
8-2 10 mm Cinegon
8-3 Inverted telephoto for camera
9-1 100 degree FOV camera lens

9-2   120 degree f/2 projection
9-3   160 degree f/2 projection
9-4   170 degree f1.8 camera lens
9-5   210 degree projection
9-6   Panoramic camera
10-2 10× eyepiece
10-3 10× eyepiece
10-4 Plossl eyepiece
10-5 Erfle eyepiece
10-6 25 mm eyepiece
11-1 10× microscope objective
11-2 20× microscope objective
11-3 4 mm apochromatic microscope objective
11-4 UV reflecting microscope objective
11-5 98× oil immersion
12-3 Flat port, 70 mm camera lens
12-5 Water-dome corrector
12-6 In-water lens with dome
13-1 5× laser beam expander
13-2 5× Gallean beam expander
13-3 50× laser beam expander
13-4 4× binocular
13-5 Power changer
13-6 Albada view finder
13-7 Door scope
14-1 Unit power relay
14-2 Unit power copy lens
14-3 0.6× copy lens
14-4 Rifle sight
14-5 Eyepiece relay
14-6 1:5 relay
15-2 Cassegrain lens 3.2–4.2 micron
15-3 Starlight scope objective f/1.57
15-4 1000 mm focal length Cassegrain
15-5 50 inch focal length telescope objective
15-6 10 inch fl f/1.22 Cassegrain
15-7 Schmidt objective

List of Lens Designs

15-8 Reflecting objective
15-9 250 inch, f/10 Cassegrain
15-10 Ritchey–Chretien
16-1 25 mm focal length periscope
16-2 65 mm format periscope
17-2 FLIR lens system
17-3 IR lens, 3.2 to 4.2 micron
17-5 10× beam expander
18-1 UV lens quartz, CaF_2
18-2 Cassegrain objective, all quartz
18-3 Lithograph projection
19-1 Document scanner
19-2 Argon laser scanner lens
19-3 Scan lens 0.6328 micron
20-1 f/3 endoscope
21-1 65 mm f/4 10× enlarging lens
22-1 f/1.8 projection lens
22-3 Projection lens, 70 mm film
22-4 70 degree FOV projection lens, f/2
22-5 Plastic projection lens
22-7 2 inch FL LCD projection
23-1 Profile projector, 20×
23-2 Telecentric lens f/2.8
23-3 f/2 telecentric
24-1 Video disk f/1
24-3 Laser focusing lens 0.308 μ
24-4 Laser focus lens 0.6328 μ
25-1 Heads up display
25-2 Biocular lens
27-4 Lens for rotary prism camera
28-1 2× anamorphic attachment, 35 mm film
28-2 2× anamorphic attachment, 70 mm film
28-3 1.5× anamorphic expander
29-1 Fused quartz condenser, 1×
29-2 Fused quartz condenser 0.2×
29-3 Pyrex condenser
29-4 Condenser system for 10× projection

30-1 5 inch f/4 aerial camera lens
30-2 12 inch f/4 aerial lens
30-3 18 inch f/3 aerial lens
30-4 24 inch f/6 aerial lens
31-1 Radiation-resistant lens 25 mm f/2.8
32-1 24× microprojection
35-1 10× zoom lens
35-2 Afocal zoom for microscope
35-3 Zoom Cassegrain
35-4 Zoom rifle scope
35-5 Stereo zoom microscope
35-6 Zoom microscope
35-7 20 to 110 mm zoom lens
35-8 25 to 125 mm with three moving groups
35-9 TV zoom lens
36-1 Optical zoom, 100 to 200 mm EFL
36-2 SLR optical zoom 72 to 145 mm EFL
36-3 6.25 to 12.5× optically compensated projection lens
37-1 Xerographic zoom
37-2 Copy lens
38-1 Variable focal length projection for SLR camera
38-2 Variable focal length motion picture projection lens
39-3 10× microscope objective, axial gradient
39-4 Laser focusing lens, axial gradient
39-5 Radial gradient, 50 inch focal length
39-6 Selfoc lens

1
The Method of Lens Design

Given an object and image distance, the wavelength region, and the degree of correction for the optical system, it would at first appear that with the great progress in computers and applied mathematics it would be possible analytically to determine the radius, thickness, and other constants for an optical system. Neglecting very primitive systems, such as a one- or two-mirror system or a single-element lens, such a technique is presently not possible.

The systematic method of lens design in use today is an iterative technique. By this procedure, based upon the experience of the designer, a basic lens type is first picked. A paraxial thin lens and then a thick lens solution is next developed. In the early days of computer optimization, the next step was often correcting for third-order aberrations (Hopkins et al., 1955). Now, with the relatively fast speed of ray tracing, this step is generally skipped and one goes directly to optimization by ray trace.

In any automatic (these are really semiautomatic programs because the designer must still exercise control) computer optimization program, there must be a single number that represents the quality of the lens. Since the concept of a good lens vs. a bad lens is always open to discussion, there are thus several techniques for creating this merit function (Brixner, 1978). The ideal situation is a merit function that considers the boundary conditions for the lens as well as the image defects. These boundary conditions include such items as maintaining the effective focal length (or magnification), f#, center and edge spacings, overall length, pupil location, element diameters, location on the glass map, paraxial angle controls, and paraxial height controls. There also should be a means of changing the weights of these defects so that the axial image quality can be weighted differently from the off-axis image as well as changing the basic structure of the image (core vs. flare, distortion, chromatic errors, etc.) (Palmer, 1971).

As an example of a merit function, let d be a defect item, the departure on the image surface between the traced rays and its idealistic or Gaussian value (or other means for determining the center of the image). These defect items we will weight (W) to permit us to control the type of image we desire. If there are N defect items, then

$$\text{Merit function} = \Sigma_{i=1}^{i=n} W_i d_i^2$$

Most merit functions are really "demerit functions," which represent the sum of the squares of various image errors: the larger the number, the worse the image. The input system is ray traced, and the merit function is computed. One of the permitted parameters is then changed, and a new value of the merit function is calculated. A table is then created of merit function changes vs. parameter changes. Then usually by a technique of damped least squares (or sometimes by orthonormalization of aberrations) an improved system is created. There are four important characteristics about this process that one should keep in mind.

    1. The process finds a local minimum. That is, a local minimum is reached in respect to a multidimensional plot of all the permitted variables. Only by experience can we really be sure that we have found a global minimum. A trend in optimization programs is the inclusion of a feature in which the program tries to find this global minimum. It does this by making many perturbations of the original system.

The orthonormalization process appears sometimes to penetrate these potential barriers. A good procedure then is first to optimize in damped least squares mode and then to orthonormalize. Another technique is to make an arbitrary change in the system and then optimize. This sometimes gets you out of the local minimum and one hopes into a better solution region.

    2. The process finds an improvement, regardless of how small, wherever it can. Thus if you do not carefully provide bounds on lens thickness, thin lenses of 1/50 lens diameter or thick lenses (to solve

for Petzval sum) of 12 inches could result. Likewise when glasses are varied, a very small improvement may result in a glass that is very expensive, is not readily available, or has undesirable stain or transmission characteristics. These computer programs can trace rays thru lenses with negative lens thicknesses. Or the lens system may be so long that it cannot fit into the "box" that has been allocated to it. So a carefully thought out boundary control is vital.

3. Computer time is proportional to the product of the number of rays traced and the number of parameters being varied. The inexperienced designer feels that he gets a better lens if he traces more rays. Instead all he gets is longer computer runs. The ideal situation is to trace the minimum number of rays in the early stages. Only increase the number of rays at the end.

4. The program neither adds nor subtracts elements. So if we start with a six-element lens, it will always be a six-element lens. It is the art of lens design to know when to add or remove elements.

The use of sine wave response considering diffraction effects (MTF) is now common in all lens evaluations. The main difficulty in using a sine wave response as a means of forming a merit function is the very large number of rays that need to be traced as well as the additional computation necessary to evaluate sine wave response. This would result in an excessively long computational run. The net result of this is that diffraction-based criteria, particularly sine wave response, háve not been used as a means of optimizing a lens but have been limited to lens evaluation.

OPTIMIZATION METHODS

In the least squares method, the above merit function is differentiated with respect to the independent variables (construction parameters) and equated to zero. The derivative is determined by actually incrementing a parameter and noting the change in the merit function. This results in N equations of the parameter increments. Some sort of matrix method for solving these equations for the parameter increments is employed (Merion, 1959; Rosen and Eldert, 1954).

It has long been known that the least squares method suffers from very slow convergence (Feder, 1957). In order to speed convergence, the concept of a metric (M) is introduced (Lavi and Vogl, 1966, p. 15). We may oversimplify and say that the gradient obtained from a least squares technique is multiplied by M to speed convergence. However, this step length computed on the basis of linearity is usually too large, causing the process to oscillate. One then introduces into M a damping factor that is large when the nonlinearity is large.

The above method will rapidly improve a crude design. After a while, a balance of aberrations will be reached. These residual aberrations vary only slowly when their construction parameters are changed (Grey, 1963). Thus the construction parameters have to be given an infinitesimal increment, which of course reduces the rate of convergence of the merit function. To avoid this problem, one must consider the rate of change of each of the defect items with respect to the construction parameters. The main difficulty in any automatic differential correction method lies not in that optical systems are nonlinear, and we use linear predictions, but in that every construction parameter affects every defect item.

In the orthonormal method, we construct a set of parameters that are orthonormal to the construction parameters. This transformation matrix relating the classical aberrations to their orthonormal counterparts is constructed at the beginning of each pass. The merit function then is expressed as the sum of squares of certain quantities, each of which is a linear combination of the classical aberration coefficients. These are orthonormal aberration coefficients because the reduction of any one of these reduces the value of the merit function no matter what the other coefficients may be (Unvala, 1966).

RAY PATTERN

Since a ray may be regarded as the centroid of an energy bundle, it is convenient to divide the entrance pupil into equal areas and to place a ray in the center of each area. For a centered optical system, one need only trace in one half of the entrance pupil. Likewise for an axial object, only one quadrant need be traced. For systems that lack symmetry, the full pupil needs to be traced.

When starting the design, always trace the minimum number of rays. A reasonable value might be.

The Method of Lens Design

Entrance Pupil Fractions					
Axial	0.866	0.5			
Off-axis tangential	0.866	0.5	0.0	–0.5	–0.866
Off-axis sagittal	0.6	0.9			

Table 1-1 Entrance Pupil Fractions, Based on Equal Areas

\multicolumn{5}{c}{Number of Rays}

2	3	4	5	6
0.866	0.913	0.935	0.948	0.957
0.500	0.707	0.791	0.837	0.866
	0.408	0.612	0.707	0.764
		0.353	0.548	0.645
			0.316	0.500
				0.289

As the design progresses, carefully monitor the ray intercept plots. If there is considerable flare, then add additional rays. Likewise, if there are problems in the skew orientation, some skew rays should be added.

In a similar manner, the field angles should be such as to divide the image into equal areas. The image height fractions are (the first field angle is axial, $N=1$)

$$H(J) = \sqrt{\frac{J}{N-1}}$$

N	2	3	4	5	6
H(1)	1.0	0.7071	0.5774	0.5	0.4472
H(2)		1.0	0.8165	0.7071	0.6325
H(3)			1.0	0.8660	0.7746
H(4)				1.0	0.8944
H(5)					1.0

ASPHERIC SURFACES

Most modern computer programs have the ability to handle aspheric surfaces. For mathematical convenience, surfaces are generally divided into three classes: spheres, conic sections, and general aspheric. The aspheric is usually represented as a tenth-order polynomial. Let X be the surface sag, Y the ray height, and C the curvature of the surface at the optical axis. Then

$$X = \frac{CY^2}{1+\sqrt{1-Y^2C^2(1+A_2)}} + A_4Y^4 + A_6Y^6 + A_8Y^8 + A_{10}Y^{10}$$

This represents the surface as a deviation from a conic section. A_2 is the conic coefficient, $= -\varepsilon^2$, where ε is the eccentricity as given in most geometry texts.

A_2 = zero   sphere
$A_2 < -1$ hyperbola
$A_2 = -1$ parabola
$-1 < A_2 < 0$ ellipse with foci on the optical axis
$A_2 > 0$ ellipse with foci on a line normal to optic axis

With present technology, it is possible to turn an aspheric surface with single point tooling. This is done with a complex numerical control system. It is being increasingly used for long-wavelength infrared systems. In the visual and UV regions, aspherics must be individually polished. The problem is twofold:

1. Most optical polishing machines have motions that tend to generate a spherical surface.
2. Aspheric surfaces are very difficult to test.

The author's best advice concerning aspherics is that unless you have to, do not be tempted to use an aspheric surface. Of course if the lens is to be injection molded, then an aspheric surface is a practical possibility. This is often done in the case of video disc lenses.

The Method of Lens Design

As an aid in manufacturing and testing aspheric surfaces, the author has written a computer program to calculate the surface coordinates as well as the coordinates of a cutter to generate this surface. Let the aspheric surface have coordinates X and Y and be generated by a cutter of diameter D. The coordinates of the center of this cutter are U and V. The cutter is always tangent to the aspheric surface. Then referring to Figure 1-1, we obtain

$$\tan \phi = | \frac{2CY}{1 = \sqrt{1 - C^2 Y^2 (1 + A_2)}} + \frac{(1 + A_2) C^3 Y^3}{\left[1 + \sqrt{1 - C^2 Y^2 (1 + A_2)} \right]^2 \sqrt{1 - C^2 Y^2 (1 + A_2)}}$$

$$+ 4A_4 Y^3 + 6A_6 Y^5 + 8A_8 Y^7 + 10A_{10} Y^9 |$$

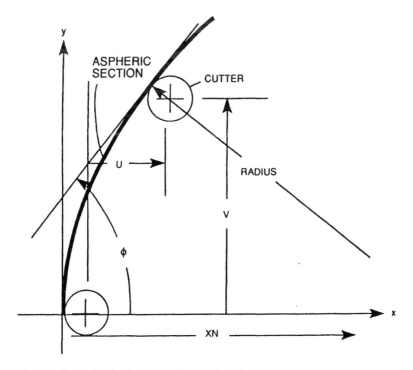

Figure 1-1 Aspheric generation and testing.

$$X = \frac{CY^2}{1+\sqrt{1-Y^2C^2(1+A_2)}} + A_4Y^4 + A_6Y^6 + A_8Y^8 + A_{10}Y^{10}$$

$$U = X + 0.5D \cos\phi - 0.5D \qquad V = Y - 0.5D \sin\phi$$

$$XN = X + \frac{Y}{\tan\phi} \qquad \text{radius} = \sqrt{Y^2 + (XN-X)^2}$$

where XN is the radius of curvature of the aspheric surface at the optical axis (the paraxial radius of curvature). See the program Cutter in Appendix E

REFRACTIVE INDEX CALCULATIONS

The spectral region of interest should be divided so as to achieve nearly equal refractive index increments. Due to the manner in which the refractive index varies for typical optical materials, it is preferable to divide the spectral region into equal frequency regions rather than by wavelength. That is, it is divided into nearly equal reciprocal wavelength increments.

For MTF calculations, five wavelengths are used. For the various regions, as used in this text, they are (wavelengths are given in microns):

Table 1-2 Wavelengths and Weights

Weight	Visual	UV	3.2–4.2 μm	8–14 μm
0.3	0.48	0.2	3.2	8.0
0.6	0.51	0.23	3.4	8.96
1.0	0.546	0.27	3.63	10.2
0.6	0.59	0.32	3.89	11.8
0.3	0.644	0.4	4.2	14.

Glass catalogs contain index of refraction values only at the various spectral and selected laser lines. Calculation at an arbitrary wavelength is performed by using a six-term interpolation formula. The coefficients for this formula for the various glasses are given in the glass catalog. This author (like most designers and commercially available computer optimization programs) has the coefficients for the entire glass catalog as well as various other optical materials, in this computer. It only remains then to type in the desired wavelengths. A program (Refractive Index) to perform this as well as the coefficients for a variety of glasses and other materials is given in Appendix E.

A typical interpolation formula is

$$N^2 = F_1 + F_2\lambda^2 + F_3\lambda^{-2} + F_4\lambda^{-4} + F_5\lambda^{-6} + F_6\lambda^{-8}$$

where $F_1, F_2 \ldots F_6$ are the coefficients for this particular glass and λ is the wavelength in microns. Another formula is the Sellmeier equation (Tatian, 1984):

$$N^2 - 1 = \frac{F_1\lambda^2}{\lambda^2 - F_4} + \frac{F_2\lambda^2}{\lambda^2 - F_5} + \frac{F_3\lambda^2}{\lambda^2 - F_6}$$

where the coefficients F_4, F_5, and F_6 represent absorption bands for that particular material.

GLASS VARIATION

Most modern lens optimization computer programs have the ability to vary the index and dispersion of the material. This assumes a continuum of the so-called glass map. This then generally precludes this variation in the UV or infrared regions. However, in the visual region, it is a very powerful variable and should be utilized wherever possible.

To be effective at glass variation, the computer program must be able to bound the glasses to the actual regions of the map. That is, if the refractive index left as an unfettered variable, we will soon have a prescription with refractive index values of 10.

But not only do you have to carefully bound values of refractive index and dispersion, the designer has to be careful as to the glass he chooses. Consider for example LAKN7 vs. SK15 or the recently reformulated N-SK15 (which are fairly close to each other on the glass map). The former, in grade A slab, costs $79/pound, while the latter costs $38 (1998 price). So for a large diameter lens, we have greatly increased the price. Of course, selecting a LASF type glass can cost from $384/pound (LASF3) to $810/pound (LASFN31) also for grade A slab.

Price is only the beginning. The designer must also check the catalog for

Availability. Some glasses are more available than others. These so-called preferred glasses are indicated in the catalog. Due to environmental pressures, particularly in Europe and Japan, some glasses have been discontinued (like LAK6) while others have been reformulated. Lead oxide is being substituted with titanium oxide. This reduces the density, and these glasses have nearly the same index and dispersion as their predecessors (SF6 and SFL6). Likewise, arsenic oxide and cadmium have been eliminated.

Transmission. Some glasses are very yellow, particularly the dense flints. This is due to the lead oxide content, with the new versions of these glasses, it is worse. For example, the old SF6 containing lead oxide has a transmission of 73% for a 25 mm thickness at 0.4 microns, while SFL6 has a transmission of 67%. Catalogs give transmission values at the various wavelengths. In the so called minicatalog, a value for transmission at 0.4 microns through 25 mm path is given.

Staining and weathering. Glass is affected in various ways when contacted by aqueous solutions. Under certain conditions, the glass may be leached. At first, when thin, it forms an interference coating. As it thickens, it slowly turns white. Interactions with aqueous solutions, particularly during the polishing operations, may cause surface staining. Glasses that are particularly susceptible are listed in these glass catalogs. Glasses that are susceptible to water vapor in the air (listed as climatic resistance) should never be used as an exterior lens elements.

Bubbles. Some glasses, because of their chemical compositions, are prone to having small bubbles. Thus these glasses cannot be used near an image surface.

Striae. A few glass types are prone to fine striae (index of refraction variations). These glasses should not be used in prisms or in thick lenses.

When the glass is finally selected, the actual catalog values are then substituted for the "fictitious glass" values. This is done by changing the surface curvatures to maintain surface powers. Let Φ be the surface power at the Jth surface with curvature C and a fictitious refractive index N. Then

$$\Phi = (N_J - N_{J-1})\, C_J = (N'_J - N'_{J-1})\, C'_J$$

N' is the catalog value of the refractive index and C'' is the adjusted value of the curvature.

CEMENTED SURFACES

Cement thickness is generally less than 0.001 inch So this cement layer is generally ignored in the lens design process. Modern cements can withstand temperature extremes from –62°C to greater than 100°C (Summers, 1991, Norland, 1999).Since these cements have an index of refraction of about 1.55, there will be some reflection loss at the interface. Nevertheless, the cement–glass interface is rarely antireflection coated. For certain critical applications, a λ/4 coating at each glass surface prior to cementing will greatly reduce this reflection loss (Willey, 1990).

Due to transmission problems with cements, their use is limited to the visual region.

ANTIREFLECTION FILMS

For light striking at normal incidence on an uncoated surface, the reflectivity R is given by

$$R = \left[\frac{N_0 - N_1}{N_0 - N_1} \right]^2$$

where light is in media N_0 and is reflected from media of index of refraction N_1. For air, $N_0 = 1$, and if $N_1 = 1.5$ then $R=4\%$.

One of the earliest antireflection coatings was a single layer of λ/4 optical thickness of magnesium fluoride. This material has an index of refraction of 1.38 (at λ = 0.55 μ).

For zero reflectivity (at one wavelength) the coating should have an optical thickness of λ/4 and have a refractive index of $\sqrt{N_s}$ where N_s is the refractive index of the substrate.

For magnesium fluoride and LASFN31, N_e = 1.88577, this is an ideal antireflection coating (see curve A in Fig. 1-2; compare it to curve D for BK7).

Consider a two-layer coating (a V coat) useful for laser systems and devices where there is only one wavelength to consider. For zero reflectivity, $N_1^2 N_s = N_2^2 N_m$.

Curve B in Fig. 1-2 shows such a V coating using magnesium fluoride as N_1 and Al_2O_3 as N_2 on BK7 glass (N_s).

To obtain a very low reflectivity over the visual region a three-layer coating is required. This has less than 0.5 % reflectivity over the region. Such a coating can be achieved with

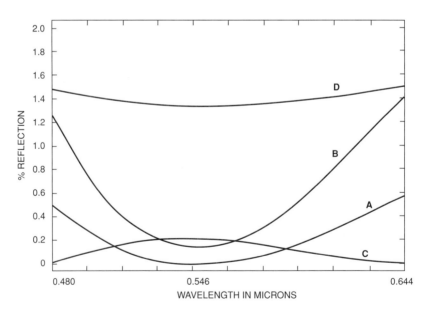

Figure 1-2 Antireflection coatings.

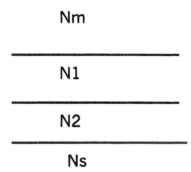

Figure 1-3 A two-layer antireflection coating.

$\lambda/4$ of MgF_2 as the first coating
$\lambda/2$ of ZrO_2 as the middle coating
$\lambda/4$ of CeF_3 next to the substrate

Curve C in Fig. 1-2 shows such coating on BK7 glass.

It is generally not necessary for the lens designed to specify the details of the antireflection coating. All that is necessary is the wavelength region and the maximum reflectivity. This is because all optical shops have their own proprietary coating formulas.

VIGNETTING AND PUPIL SHIFT

Vignetting is a reduction in the size of the entrance pupil, for off-axis objects, due to physical constraints of lens diameters. By this definition there is no vignetting on-axis. The size of the axial entrance pupil is determined by the system f# or numeric aperture.

Let us define the entrance pupil for off-axis objects as perpendicular to the chief ray. Since the entrance pupil is an aberrated image of the aperture stop, our first step is to determine the aperture stop diameter by tracing an upper rim axial ray. Then for this off-axis object, we determine the size of the vignetted entrance pupil by iteratively tracing upper, lower, and chief rays to the aperture stop. For certain systems, the full, unvignetted entrance pupil may not be traceable. We therefore apply this vignetting and pupil shift to the aperture stop.

That is, the ray coordinate data is shifted and vignetted onto the aperture stop.

Vignetting is nearly the same for all wavelengths (at any particular field angle). So for convenience, we need only to do the above at the central wavelength. However, there is in general a different vignetting and pupil shift for all off-axis field points and configurations (a zoom lens). In a typical computer program, vignetting and pupil shift are handled as follows.

VTT(J) is the vignetting for a particular field and configuration. It is expressed as a fraction of the aberrated entrance pupil diameter. It applies only in the meridional direction. See Fig. 1-4.

VTS(J) is the same but in the sagital direction. For a centered optical system with rotational symmetry, these items are one.

PST(J) is pupil shift for the vignetted pupil corresponding to the above.

PSS(J) is the same but in the saggital direction. For a centered optical system with rotational symmetry, these items are zero.

In the absence of any pupil shift or vignetting, all VTT(J) and VTS(S) items are set to 1. and all PST(J) and PSS(J) are set to 0.

To trace to the top of the pupil, PST(J) + VTT(J) = 1.

Entrance pupil coordinates are then shifted and multiplied by the appropriate vignetting coefficient. This applies to all raytracing, MTF data, spot diagram, lens drawings, etc. By this technique, accuracy in MTF and spot diagram computations is not compromised, since the full number of rays is being traced in the presence of vignetting. It also

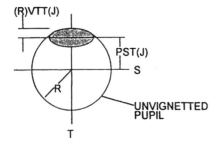

Figure 1-4 The vignetted entrance pupil.

allows the designer deliberately to introduce vignetting into the system when it is necessary to constrain lens diameters.

The lens designer must be cautioned that in systems with considerable vignetting, care must be taken that there are lens diameters to limit the upper and lower rim rays of the vignetted pupil. The aperture stop is now not the limiting surface. Only rays that were traced in optimization must be able to pass through the optical system.

CHANGING THE NUMBER OF ELEMENTS

Sometimes during the course of the design, the designer will note that an element is becoming very thin and of very low power. In this case the program's curvature bounds and thickness bounds are invoked to make it a nearly zero power element. Then the element is removed from the prescription.

A tougher case is when the lens image quality is not adequate. The usual advice is to add an element. But where should such an element be added, and how? There are several choices.

1. In front of or behind the lens. This is easy: just add a plane plate, guessing at the material type. First vary only the curvatures of this new element along with the curvature and thickness parameters of the remaining elements. Then vary the index and dispersion of the new element. (This obviously is not possible in the infrared or UV regions.)

2. Insert a plate in a large air space in the lens. Remember to readjust the air spaces such that if D was the original air space and D_1 and D_2 are the new air spaces and the plate has a thickness T and refractive index N, then

$$D = D_1 + D_2 + \frac{T}{N}$$

3. Split a very thick lens into a cemented doublet. This might be a logical choice if there is considerable chromatic aberration. Vary the refractive index and dispersion of both new elements.

4. Split a very thick lens into two elements separated by a very small air space. This could lead to ray trace difficulties in regions of large angles of incidence. Keep in mind that with today's coating technology, the cost of two air-spaced elements is nearly the same as that

of a cemented doublet. The splitting is generally done with two plane surfaces, separated by a small air gap.

VARIABLE PARAMETERS

The beginning designer often asks, "What lens parameters should I vary?" The answer of course is that one should vary them all—but not right away. My general procedure is as follows.

First Series of Runs

Vary all the radii, large air spaces, and positive lens thicknesses. If there is an aspheric surface, then these coefficients should be varied. Radii and aspheric coefficients represent the most powerful lens parameters and so should be varied from the very beginning of the design phase. Positive lens thicknesses should be varied because it is necessary to control lens edge thickness. Review all the bounds to be sure that the lens is buildable and that it will fit your requirements as to diameter, overall length, back focus, etc.

Second Series of Runs

Add to the above parameters the thicknesses of negative lenses as well as the remaining air spaces. If some of these spacings are giving problems (going to their maximum or minimum values), it is best just to fix this parameter (at least for the time being).

Third Series of Runs

Add to the above parameters the index and dispersion of glasses. Obviously this is skipped if in the UV or IR regions. Then fix the glasses. The author finds this the most soul-searching part of lens design, since one must now make value decisions regarding glass prices and availability, stain and bubble codes, and of course performance.

Fourth Series of Runs

With the glasses fixed, again vary all the parameters (except obviously index and dispersion).

The Method of Lens Design 17

At several stages in the design process, it is wise to

1. Run MTF calculations to be sure that the design is meeting your image quality requirements.
2. Check distortion.
3. Plot the lens to be sure that it is buildable.
4. Examine the intercept and path length error plots as well as third-order surface aberration contributions. This often gives the designer an insight into his design problems. Based upon this, he might want to split a lens, add a lens, etc. (For a discussion on third-order aberrations, see Born and Wolf, 1965.)

BOUNDS ON EDGE AND CENTER THICKNESS

For the economical production of a lens element,

> Negative elements should have a diameter-to-center thickness ratio of less than 10. This ratio is necessary to prevent the lens from distorting when removed from the polishing block. Diameter-to-thickness ratios as high as 30 are of course possible, but production costs increase. In the IR region, where material is expensive and there is considerable absorption and scattering, high thickness ratios are common.
>
> Positive elements should have at least 0.04 inch edge thickness on small lenses, less than one inch in diameter, and at least 0.06 inch edge thickness for larger lenses. This is necessary to prevent the lens from chipping while being processed.

TEST GLASS FITTING

During the polishing process, all spherical surfaces are compared under a monochromatic light source with a test glass (Malacara, 1978, p. 14). A test glass consists of a pair of concave/convex spheres, generally made from Pyrex or sometimes fused quartz. When compared to the work in process, Newton rings are seen, which represent contours of half-wavelength deviations the work is from a sphere. This is a crude but practical way to determine the accuracy and irregularity of the work. 1/4 and 1/8 fringe deviations are readily discernible.

Of course this technique has two disadvantages:

1. The work surface is contacted by the test glass and so may be scratched. Today, with the availability of the HeNe laser, various interferometers are available (Zygo for example) in which no surface contact is made.
2. For every value of the radius, one needs a pair of test glasses.

Every optical shop has an extensive test glass inventory. These lists are made available to the designer in the hope that he will select values of radii from this list. The cost of a test glass is approximately $400 per radius value, so the total cost of a prototype optic is vastly reduced if the designer can fit his design to the optical shop's list. Unfortunately each shop has a different list; there is no standardized list.

As a basis for such a list, one might consider a system in which each radius is a constant multiplier of the next smallest value, that is,

$$R_j = c\, R_{j-1}$$

for 100 values, between 1 and 10; $c = 1.02329$.

But a rationalized system will never happen, and so the designer has to contend with fitting his design to the irrational values of test glasses that his shop has. This author uses a rather crude technique, as follows.

1. The third-order aberration contributions at each surface are scanned, and a radius tolerance is estimated.
2. Any surface that lies within this tolerance limit of a test glass value is then actually set to the test glass value.
3. The system is then optimized, of course keeping the test glass fitted surfaces constant. (The other radii and thicknesses are varied.)
4. Steps 2 and 3 are then repeated. Values of radii that at first did not lie within its tolerance for a test glass will often move to a new value with the subsequent optimization and now can be fitted.

The author has written a computer program that reads the lens prescription data and compares the radii to the test glass list. See the program Test Glass Fitting in Appendix E. Other designers have advocated a different technique. They try to fit the most sensitive radii first. The least sensitive radii they feel can always be fitted.

Regardless of the method you choose, do not worry if you cannot fit every radius to the test list. If you fit most of them, you have still saved your client a substantial sum.

MELT DATA FITTING

Some lenses, particularly long-focal-length high-resolution types, are sensitive to small changes in refractive index and dispersion of the actual material used as compared to the nominal or catalog value. For materials such as quartz (fused silica), calcium fluoride, silicon, germanium, etc., the refractive index is an intrinsic property of the compound. For mixtures such as optical glass, there are slight variations in refractive index from batch to batch. The refractive index is carefully controlled by the glass manufacturer. Typical tolerances for glass as supplied are

$$N_d \pm 0.001 \qquad V_d \pm 0.8\%$$

The glass manufacturer generally supplies to the optical shop a melt data sheet for each supplied batch of glass. These sheets contain the actual measured refractive indices at several spectral lines for that particular batch of glass. In the event that the measured values depart by more than the tolerance limit for the lens, then the values of radii, lens thicknesses, or air spaces need to be adjusted. This process is called melt data fitting and fortunately needs to be performed only on a few types of lenses.

The process becomes complex if the lens designer used refractive index values at wavelengths other than those that the glass manufacturer measured. For example, if data is supplied at the spectral lines e, f, c, and g but the designer requires data at 0.52 microns, then some sort of interpolation technique is needed. A method that seems very effective (private communication from D. Grey) is to fit by least squares the difference between the melt index of refraction and the corresponding calculated value to an equation of the form

$$R(\lambda) = A \lambda^{-3} + B \lambda^{-2} + C\lambda^{-1} + D$$

This value of $R(\lambda)$ is then added to the calculated value of the refractive index. (This calculated value is computed from the polynomial coefficients given in the glass catalog as discussed earlier.)

THERMAL PROBLEMS

Thermal problems can be divided into two classes:

1. The entire lens or mirror has been raised (or lowered) to a uniform temperature.
2. There are temperature gradients across the lens or mirror.

If an optical system has its temperature uniformly changed, then the main problem is a shift in the image surface location. The effective focal length of course changes, and there will be some loss in image quality. These changes can be reduced by various thermal compensation techniques.

In Cassegrain systems, a favorite technique is to control the spacing between primary and secondary mirrors with Invar rods. Although the entire system is in an aluminum housing, whose dimensions change with temperature, the most critical spacing, that of the primary to the secondary mirror, is now held constant with temperature.

Another technique is to make the entire system out of the same material. For an all-mirror system, this can be conveniently done, since metal mirror fabrication is now commonplace. Aluminum mirrors are often made by roughly shaping the mirror, chemically depositing nickel, and then polishing this surface to the desired figure. It is then vacuum coated with aluminum.

For a lens system, making all the spacers out of Invar does not help, since the radii, lens thickness, and refractive index of the elements change with temperature. Fortunately, most optical glass catalogs now give the change in refractive index *dn/dt* as well as the thermal expansion coefficient α. This data is then used to create a new prescription in which all radii, thicknesses, and refractive indices have been changed as a result of the temperature change. It is a complex process, since an axial spacing change is a result of how the spacer contacts the edge of the lenses. The system is then analyzed, and if there are image quality or back focal length changes (which is likely) then one substitutes a different spacer material. For example, if two elements are spaced with an aluminum alloy (6061) spacer ($\alpha = 216 \times 10^{-7}/°C$) then this space between the elements may be reduced if brass is substituted ($\alpha = 189 \times 10^{-7}/°C$) or increased with a magnesium spacer ($\alpha = 258 \times 10^{-7}/°C$).

This unfortunately is a very tedious procedure. Computer programs have been written to perform these thermal perturbations.

In the second case where there are temperature gradients, the lens, which formerly had rotational symmetry, is deformed and so lacks this symmetry. There is very little the lens designer can do about this except to use fused quartz where possible and a very low expansion material for mirrors like titanium silicate (Corning 7971; see Appendix C). The change in optical path length, resulting from a temperature variation ΔT is (Reitmayer and Schroder, 1975)

$$\Delta W = d\, [\alpha\, (n-1) + dn/dt]\, \Delta T$$

where n = index of refraction, d = thickness of the element, and α = coefficient of thermal expansion.

Unfortunately, for nearly all materials, dn/dt is a positive number; that is, the refractive index increases with temperature. There are a few materials that have a negative dn/dt. These are the FK series of glasses, PK53, PK54, SK51, LAKN 12, and LAKN13.

OPTICAL TOLERANCES

Perhaps the most neglected portion of the lens design process is the tolerancing and subsequent drawing preparation. Conceivably, this is because it is the least creative portion of the task. However, without proper tolerancing and proper drawings all of the work of the lens design process may produce an inferior or even an unacceptable product.

Perhaps the simplest method is the use of the merit function from the lens optimization program. That is, if the merit function as constructed is adequate for optimization, then why should it not be used to tolerance the lens? For tolerances on curvature, thickness, refractive index, and dispersion, this is a simple task. One makes a series of computer runs in which these parameters are changed by small amounts. Then by estimating a permitted increase in the merit function, one can arrive at the tolerances for the above parameters.

However, this technique becomes complex when one tries to introduce tilts, decentrations, and surface irregularities into the lens system. There are several additional considerations when tolerancing a lens system:

1. Tolerances must be assigned to each parameter by some statistical method (Koch, 1978). Everything subject to manufacture will depart from the nominal design.

2. In addition to actual image quality changes as a result of manufacture, we often must maintain certain first order parameters: effective focal length (or magnification) and back focal length. In this regard it is helpful to have printed out a table of the variation of these first-order parameters vs. the lens parameters of curvature, thickness, and refractive index.

3. There is often a parameter that may be used to compensate for image or first-order errors. The simplest case is a variation in back focal length. This is often compensated by adjusting the mounting flange as the last step in manufacture. Also in telephoto lenses, the large air space between the front and rear groups may be used to maintain effective focal length.

4. Accuracy represents the total number of rings that the surface deviates from the test glass. Irregularity is the difference in fringes as seen in two mutually perpindicular directions. This irregularity causes astigmatism for an axial object. In order to detect irregularity, the accuracy should not be greater than 4 to 6 times the irregularity. However, the accuracy also is related to the radius tolerance, particularly in cases where the radius is greater than 10 times the diameter. Let Y be the semidiameter and Z the sag at the surface. Then (approximately)

$$Z = \frac{Y^2}{2R}$$

The Method of Lens Design

Taking the derivative,

$$\Delta Z = -\frac{Y^2}{2R^2}\Delta R$$

As an example, for an accuracy of 4 fringes, $R=100$ mm and $Y= 5$ mm, $\Delta Z= 1.1 \times 10^{-3}$ mm.

ΔR then is 0.88 mm, which is probably greater than the radius tolerance (which might be 0.2 mm).

As discussed previously, should the lens be subject to refractive index or dispersion variations of less than catalog values, then melt data fitting is employed.

When the designer prepares the lens drawings, the actual tolerances are a blend of tolerances to maintain image quality and tolerances to maintain first-order properties.

Ginsberg (1981) discusses this overall tolerancing concept. (I would like to thank him for his review and comments on this section.)

Figure 1-5a shows a decentered lens element of focal length F and refractive index N. The optical axis contains the centers of curvatures of the lens surfaces. However, the lens is actually centered about an axis indicated as the lens center.

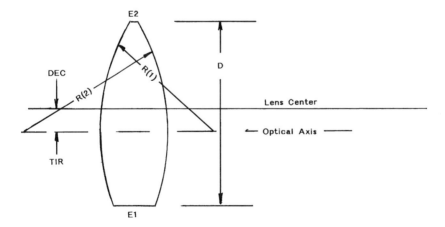

Figure 1-5a   Lens decentration.

Decenter (DEC) is the distance between the optical center and the center of the lens. Image runout is the diameter of the circle made by the nutating image as the lens is rotated. Image runout = 2 DEC.

DEC = F deviation   (See Chapter 6 of Kimmel and Parks, 1995)

$$\text{Edge variation} = E1 - E2 = \frac{D(\text{deviation})}{N-1}$$
$$= \text{TIR (total indicator runout)}$$

This assumes that the part runs true on surface 2 and rotates about the lens center axis. D is the clear aperture diameter of surface 1, and this is where the dial indicator is placed to read the TIR. As the lens is rotated, the image of a distant object will rotate about a circle of diameter 2 DEC.

Wedge (in radians) is the edge thickness difference divided by the lens diameter.

Figure 1-5b shows how a lens is set up for edging. On the left a lens is waxed to a true running spindle in preparation for edging (centering). Notice that the center of curvature of the surface in contact with the spindle lies on the center of rotation of the spindle.

In the center figure, a bright object is seen as reflected from the external surface of the lens. As the lens is rotated, the image tracks out a circle in space. With the wax still soft (a little heat might be needed) the lens is pushed closer to the centered position and the circle becomes smaller until no movement of the image can be discerned. Or a

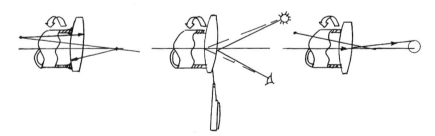

Figure 1-5b Centering (edging) an element.

The Method of Lens Design

dial indicator may be placed near the edge of the external surface. (With this method, there is the possibility of scratching the polished lens surface.)

In the figure on the right, a beam of light (perhaps a laser) is transmitted through the lens and the deviation noted. Again the lens is adjusted until no deviation can be detected.

With the lens so centered on the spindle, a diamond wheel then edges the lens to the proper diameter.

It is important to try to create a design that is not too sensitive to the construction parameters. The usual method of optimization creates a lens that is in a local minimum in respect to the "centered" parameters: thickness, curvature, refractive index, and dispersion (assuming that all these parameters were varied). However, optimization does not create a system that is a minimum in respect to tilt and decentrations. As a means of reducing these effects, the designer should try to avoid large angles of incidence and large third-order surface contributions.

In POP there is a bound (Grey, 1970, 1978) whereby sensitivity to tilt and decentration can be reduced. This, however, must be used with caution. The item bounded is the RMS OPD (optical path difference) path length error induced by tilt or decentration. (See the following discussion on wave front perturbations.) In addition, at the end of a computer run, there is a printout of lens tolerance data for each surface and for each lens. This is RMS OPD (in microns) per 0.001 inch lateral displacement as well as a TIR of 0001 inch at the edge of the clear aperture. This alerts the designer to potentially sensitive surfaces.

MTF [modulation transfer function] (or OTF [optical transfer function]) is presently regarded as the best way to evaluate an optical system. So although most merit functions are not based on diffraction MTF, final analysis of the lens is by diffraction-based MTF. Consequently, some tolerancing programs (for example Code V, Appendix D) utilize a procedure for calculating the variation of OTF with the construction parameters (Stark and Wise, 1980). In order to reduce ray trace time, one procedure (Rimmer, 1977) is to expand the OTF in a power series in the parameters of interest. Another technique is to consider wavefront perturbations as a function of parameter variations (Hopkins and Tiziani, 1966).

Let τ be the path length error produced by a perturbation.

$$\tau = (N \cos I - N' \cos R)\Omega$$

where N and N' are refractive indices; I = angle of incidence; R = angle of refraction; and Ω = motion of surface normal to ray propagation.

Consider a lens of refractive index N in air with a surface irregularity thickness of t. Then (at near normal incidence)

$$\tau = (N - 1)t$$

whereas for a mirror surface,

$$\tau = 2t$$

Now comparing a mirror system to a lens system of refractive index 1.5, the mirror is 4 times as sensitive to effects of surface irregularity as the equivalent lens. One starts by assigning tolerances uniformly, that is,

$$0.025\lambda\sqrt{M}$$

where M is the number of parameters subject to manufacturing errors. The total wave front error then will be

$$\sqrt{\sum_{i=1}^{M} T_i^2} = \frac{\lambda}{4}$$

where T_i is the wavefront error for the ith parameter. Most optical systems are not diffraction limited, and so the wave front error for each parameter may be accordingly increased.

These tolerances should be changed to reflect manufacturing charges. That is, if a lens thickness tolerance becomes ±0.020 in., it should be changed to ±.005 in. because there is no price change. This will help relieve the burden for those tolerances that become extremely tight.

Likewise the designer should consider the cost of tighter part tolerances vs. assembly, adjustment, and test time. In this regard, this author designed and had fabricated several complex lenses for undersea use. It was a motor driven lens and had an extensive assembly and test procedure. After doing several lenses, I realized that my assembly and adjustment time was excessive. By tightening many of the lens and mechanical part tolerances, I greatly reduced my assembly and adjustment time. The increase in part cost was substantially less than the cost of labor saved.

LENS DRAWINGS

Upon completion of lens tolerancing, drawings for all the lens elements are prepared. The designer should keep in mind that often drawings are prepared by a draftsman. The designer should always check that the drawings are accurate and toleranced properly. As an aid to someone else preparing the optical drawings, this author found it helpful to submit, as part of the design package, the lens prescription in a form in which all columns are clearly labeled. This is illustrated in Table 1-3 which is the lens prescription for the inverted telephoto lens of Fig. 8-1. This is also an aid to the mechanical engineer in prepara-

Table 1-3 Lens Prescription for inverted telephoto lens, f/3.5, effective focal length = 1.1811 in., f# = 3.5

	Radius	Thickness	Sum	Diameter	Sag	N	V	Edge	Volume
1	0.0000	−0.697	0.000	0.337	0.000	1.00000	0.000	−0.692	
2	16.1946	0.096	0.096	0.820	0.005	1.52232	69.499	0.201	0.089
3	0.5344	0.589	0.686	0.650	0.110	1.00000	0.000	0.650	
4	0.5774	0.118	0.804	0.820	0.171	1.72311	29.290	0.078	0.063
5	0.5313	0.270	1.074	0.700	0.132	1.00000	0.000	0.175	
6	2.3578	0.304	1.378	0.820	0.036	1.57830	41.220	0.211	0.136
7	−1.5162	0.216	1.593	0.820	−0.056	1.00000	0.000	0.272	
8	0.0000	0.074	1.668	0.640	0.000	1.00000	0.000	0.074	
9	0.0000	0.071	1.738	0.820	0.000	1.72311	29.293	0.179	0.069
10	0.7236	0.210	1.949	0.760	0.108	1.64304	59.850	0.045	0.059
11	−1.2845	2.169	4.118	0.760	−0.057	1.00000	0.000	2.226	

tion of lens spacers and housing. All data is in inches, while the lens volume (useful for weight calculations) is in cubic inches.

The distance from the first lens surface to the image is 4.118 in.

Instead of supplying *N* and *V* values, some prefer the glass code. Glass codes are presented in all the tables of lens prescriptions.

Figure 1-6 shows a typical drawing for a cemented element. It is the cemented achromat of Fig. 2-2 and is a cross-section view as accepted in the optical industry (MIL-STD-34 and ISO standard 10110), not a true view as a mechanical engineer would create.

A few comments concerning this drawing are in order. Some companies would make this into three drawings: Lens A, Lens B, and a cemented assembly drawing. However, most optical shops would just as soon deal with the single drawing. Diameter, center thickness, and radii are toleranced. Accuracy represents the total number of fringes seen when a test glass is applied to the surface. Ir-

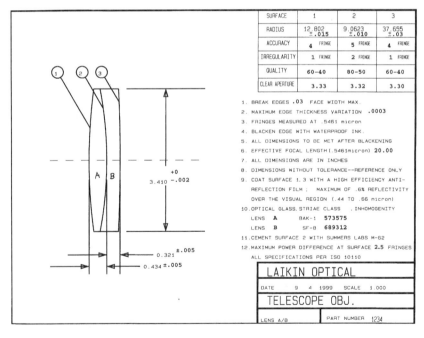

Figure 1-6 Lens drawing.

regularity is the difference in fringes between two perpendicular directions. All optical specifications apply only within the clear aperture.

With a cemented assembly, some designers prefer to have the positive element (the crown) a little smaller in diameter (perhaps 0.005–0.010) than the negative element. The smaller crown lens then can be moved during the cementing process to assure centration. The lens assembly is then located by the negative element. For this illustration, both elements were made the same diameter.

Note 1. All edges should have some break. This prevents chipping during manufacture and subsequent assembly.

Note 2. Maximum deviation in minutes of arc is sometimes given, or minutes of arc maximum wedge.

Note 3. Often lenses are used at wavelengths other than the visual region (the IR), and so it is important to define the wavelength at which the accuracy and irregularity specification apply. Also many shops are using the HeNe laser (0.6328 micron) in their testing.

Note 4. Edges are often blackened to reduce stray light (veiling glare) in the system. For very high energy laser systems, this is often deleted.

Note 5. This is to prevent any edge buildup; not necessary if an ink is used.

Note 6. EFL is a handy bit of information. It is an aid in testing the completed clement. It is reference only, and so not toleranced.

Note 7. Most shops in the United States are on the inch system. If the lens is to be made elsewhere, millimeters would probably be the correct choice.

Note 8. Reference dimensions are very convenient for test and manufacturing purposes. Sometimes lens edge thickness is indicated.

Note 9. Coatings must be specified as to maximum reflectivity over some spectral region. With modern coating technology, the single layer anti reflection coating of a quarter wavelength of magnesium fluoride is now obsolete.

Note 10. The six-digit code is per MIL-G174. The first three digits indicate refractive index while the next three indicate dispersion. For the A lens, $N_d = 1.573$ and $V_d = 57.5$. If not an optical glass (quartz, calcium fluoride, silicon, germanium, etc.), more information as to the material specification must be given. (See also ANSI, 1980, PH3.617 for an equivalent specification to MIL-O-13830.)

Note 11. This is a thermosetting cement made by Summers Labs and Norland Products.

Note 12. We want to prevent a large power difference at the cemented interface.

COMPUTER USAGE

In 1965 I was in charge of a lens design group at EOS (a division of Xerox) in Pasadena. We were using Grey's programs on a CDC 6600. Since the machine was some distance (in El Segundo) from us, we key punched our data decks and submitted this by overnight courier service. Next morning, our computer runs were brought in along with punched cards giving the new prescription. Since we could get in only one computer run per day, we carefully thought out each run.

In 1968 I had an office in a computer facility and was then self-employed. Since all I had to do was turn in a data deck to the computer operator, I was able to get in many runs each day. Although my productivity increased, my computer bill vastly increased.

Now with a personal computer at my desk I can get in many more computer runs than I had previously dreamed possible. Being older and wiser, I now see the value of carefully analyzing a computer run before submitting another one. Most lens design programs print out a lot of very useful data in addition to the lens prescription: first-order data, third-order aberration contributions, tilt and decentration sensitivity data, intercept and path length data, MTF, etc.

The designer should spend some time analyzing the results of his computer run before forging ahead and submitting another computer run. In this respect, a plot of the lens system is very helpful. Potential problems should be noted. Some of these are

   Large angles of incidence at a surface
   Very thick lenses
   Very thin lens edge
   Length of lens too long to meet system requirements
   Back focal length too short

The Method of Lens Design 31

However, there are a few times when many computer runs with very little thinking are justified. Two such cases are when one has difficulty ray tracing due to either f# or field angle. In such a situation, a method that works is to trace at the maximum aperture and field angle possible using as variables all the curvatures and most of the air spaces (the large ones) and positive lens thickness. The for the next optimization run, simply increase the field angle or aperture. In this manner, the lens is "opened up" so that it traces to maximum aperture and field.

PHOTOGRAPHIC LENSES

Following are some design considerations for photographic lenses. (Keep in mind that this represents a generalization and is indicative of photographic lenses for SLR type cameras; see also Betensky, 1980.)

1. *Distortion.* Usually less than 2%. Distortion as high as 3% is tolerable for non architectural scenes.
2. *Focus.* Since most lenses are focused at full aperture, the shift should be less than 0.02 mm when going from wide open to smallest aperture.
3. *Vignetting.* 20% can be tolerated at full aperture. There should be no vignetting at half the maximum aperture.
4. *Veiling glare.* Less than 1% good, less than 3% acceptable, less than 6% poor.
5. *Spectral region.* Due to the blue sensitivity of most films it is important to trace to the g line (0.4358 micron); the center could be 0.52 micron with the long wavelength at C' (0.6438 micron). However, lenses for the graphic arts industries are generally used with orthochromatic emulsion films (Kodak #2556). A good choice would be h (0.4047 μ), F' (0.48 μ), and e (0.5461 μ).

Although MTF data for all "visual" lenses (as used in this text) have weights and wavelengths as given previously (see Table 1-2), a better choice, for strictly photographic use (considering its blue sensitivity), would be values as given in Table 1-4 (Betensky, 1980).

Table 1-4 Wavelength vs. Weight for Photographic Lens

Wavelength (μ)	Weight
0.4358	0.3
0.474	0.6
0.52	1
0.575	0.6
0.6438	0.3

6. *Iris.* All photographic lenses are fitted with an iris (in contrast to projection lenses, which have no iris). Sufficient clearance must be allowed, generally about 0.12 in. on both sides.

LENSES FOR USE WITH TV TYPE SYSTEMS

These detectors are generally vidicon or charge coupled devices. Since their spectral response curves are often different from a "visual" or photographic system, the designer should adjust the wavelength region accordingly.

Resolution for these types of systems are generally lower than for a photographic system. Also note that in electronic data sheets, resolution is often expressed as TV lines. This represents the actual number of scan lines, not line pairs/mm as in optical references. For example, consider a typical one-inch vidicon (so called because the outside portion of the tube is one inch in diameter. It has a vertical height of 0.375 in. At 525 TV lines its resolution would be

$$\frac{525}{(0.375)2} \text{ lp/inch} = 28 \text{ lp/mm}$$

So we want high MTF response at low spatial frequencies. This is accomplished by trying to reduce image flare. We then adjust our ray pattern to trace more rays at the outer portion of the pupil.

The Method of Lens Design

Sometimes it is required to have a system with a large dynamic range. Such lenses are generally fitted with an iris. Sometimes this iris is coupled to the photocathode to maintain a constant response as the ambient is changed. Unfortunately, the minimum practical diameter for an iris is about one mm. In order to increase the dynamic range, a neutral density spot is placed in close proximity to this iris (Busby, 1972). In a typical system, this spot may only obscure 1% of the area at the aperture stop. If its transmittance is 0.5%, then the dynamic range has been increased by 200.

INFRARED SYSTEMS

Several computer programs consider both intercept error and path length errors in their merit functions. These errors are weighted to strike a balance between these types of errors. However, such a balance is generally based on visual correction. At much longer wavelengths, it is important to be able to manipulate the merit function to decrease the weight on pathlength errors.

Narcissus is an important consideration in infrared systems (see msp. 17). Following Howard (1982) and referring to Fig. 1-7; we trace

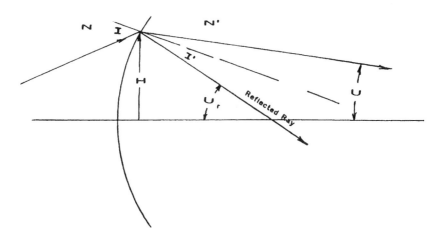

Figure 1-7 Reflected ray in narcissus.

two unrelated paraxial rays, a ray reflected from the detector and a forward ray going toward the detector. Then the Lagrange invariant Ψ is

$$\Psi = H_r N' U - H N' U_r$$

and since $H_r = H$,

$$\Psi = H N' [U - U_r] \qquad U_r = U + 2I' \qquad \Psi = -H N' 2I' = H N 2I$$

At the detector plane, $\Psi = H_r N' U'$. Thus H_r, the radius of the circular ghost at the detector plane, is

$$H_r = \frac{2 H N I}{N' U'}$$

It is helpful to have this the value *HNI* printed out, on a surface-by-surface basis, to determine if any surfaces will contribute to narcissus. This narcissus effect, the reflection of the detector back upon itself, is only of importance for surfaces prior to the scanning mirror. That is, a scanned reflection from a surface in front of the scanning mirror causes an AC signal, whereas reflections from surfaces past the scanner only cause a DC signal to be impressed upon the detector.

UV SYSTEMS

The comments above concerning pathlength errors vs. intercept errors of course apply here. In this case the weight on path length errors should be increased (see Chap. 18).

SECONDARY COLOR

Perhaps the most difficult aberration to control is secondary color. This becomes acute in long focal length, large f# systems. In a typical visual system correction, the F' and C' foci are united behind the e focus. This longitudinal distance is approximately focal length/2000. The 2000 is a consequence of glass chemistry.

The Method of Lens Design

$$V = \frac{N_e - 1}{N_{f'} - N_{C'}} \quad P = \frac{N_{f'} - N_e}{N_{f'} - N_{C'}}$$

If one plots all optical glasses for values of V vs. P, a near straight line is obtained. This is shown in Fig. 1-8, which is a plot of readily available optical glasses along with some additional optical materials (see the program Refractive Index in Appendix E). In order to reduce secondary color, Conrady (1957, p. 158) showed that it is necessary to use a material that departs from this "glass line."

Unfortunately, there are only a few materials that depart from this glass line:

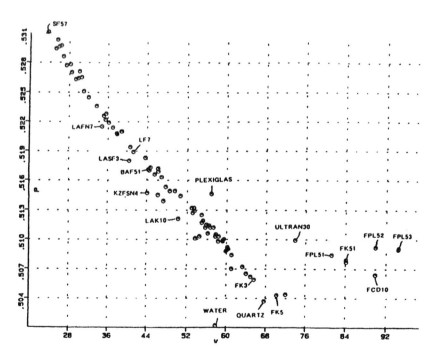

Figure 1-8 Secondary color chart.

KZFS type glasses: KZFSN4
FK type glasses: FK51 (Schott); S-FPL51, S-FPL52, and S-FPL53 (Ohara); FCD-1, FCD-10, FCD-100 (Hoya)
Calcium fluoride

Be careful when selecting some of these materials, since they are not always readily available.

To reduce secondary color with KZFSN4 glass, it should, like all materials to the left of the glass line, be a negative lens. Unfortunately, its departure from the glass line is not great enough to be very effective in secondary color reduction.

As can be seen from Fig. 1-8, some of the FK materials and calcium fluoride greatly depart from the glass line. Calcium fluoride is a very transparent cubic crystal, expensive (in comparison to optical glasses), fragile, very high thermal expansion coefficient, soft, and very slightly hydroscopic. It is used extensively in microscope objectives (fluorite) because the material costs in small diameters become insignificant. FK-51, SS-FPL51, S-FPL52, S-FPL53, FCD-10, and FCD-100 are glasses with high expansion coefficients and therefore some optical shops feel they are difficult to polish. Some of these FK materials are prone to striae. Neither of these materials should be used as an exterior element in an optical system. Because they lie to the right of the glass line, they become positive lenses when used for secondary color correction. S-FPL53 is a relatively new material and it promises to be the most effective way to reduce secondary color. Its values of V and P are very close to those of CaF_2.

THE DIFFRACTION LIMIT

For an optical system with a uniformly (incoherent) illuminated, circular entrance pupil,

$$\text{Depth of focus} = 2\lambda\, (f\#)^2 (1 + m)^2$$

$$\text{Resolution} = \frac{1000 \text{ lp/mm}}{\lambda f\# (1+m)} = \frac{1818 \text{ lp/mm}}{f\# (1+m)} \quad \text{at} \quad \lambda = 0.55 \mu m$$

$$\text{Radius of Airy disc} = 1.22 \lambda f\# (1+m)$$

where λ is wavelength in microns, m is the absolute value of the magnification (equals 0 for objects at infinity), and f# is the ratio of focal length/entrance pupil diameter.

This is the basis of the Raleigh criterion of resolution. Raleigh felt that he could resolve two objects that were an Airy disc radius apart. According to the Sparrows criterion we can do a little better than this. Using this criterion,

Least linear separation resolvable = λf#(1+m)

If the system is not diffraction limited and has a resolution of R in lp/mm, then

$$\text{Depth of focus} = \frac{2\text{f\#}(1+m)}{R}$$

DEPTH OF FIELD

If a lens is focused at a distance D then the depth of field as measured from this object point is (see Eq. 6.6 of Smith, 1966)

$$\frac{D^2 b}{A \pm bD}$$

where b is the permitted angular blur and A is the entrance pupil diameter.

$$b = \frac{B}{F(1+m)}$$

where B is spot size diameter, F the focal length, and m the absolute value of the magnification.

Let H be the hyperfocal distance. Then

$$H = \frac{F^2}{f^\# B}$$

This assumes that $H \gg F$. For the usual photographic lens B is approximately 0.025 mm. The near distance at which the lens is acceptably sharp is

$$\frac{HD}{H+D-F}$$

and the far limit is

$$\frac{HD}{H-D+F}$$

For $H \gg F$ and if $D=H$ (the hyperfocal distance) then it is acceptably sharp from $H/2$ to infinity. In the usual photographic lens, values of these near and far limits are engraved on the lens housing so that the photographer may note these limits for the various settings of his iris.

DIFFRACTION LIMITED MTF

Consider a lens with a uniformly illuminated entrance pupil in incoherent light. Then the MTF response is

$$T = \frac{2\left[\arccos(K) - K(1-K^2)^{1/2}\right]}{\pi}$$

where K is the normalized spatial frequency $= S\lambda f^{\#}$ and S is the spatial frequency in lp/mm for which one wants to find the response. K then lies between 0 and 1.

For easy computation of MTF diffraction limited response, Table 1-5 gives values of K vs. T (see program MTF in Appendix E).

Consider, for example a 50 mm f/2.8 photographic lens ($\lambda=0.55$ μ). Then

$$\frac{1}{\lambda f\#} = 649 \text{ lp/mm}$$

which is the diffraction limit, and so at $S = 100$ lp/mm, $K = 0.15$. Then referring to the table, the response is 0.81 (of course most photographic lenses are not diffraction limited at f/2.8).

The Method of Lens Design

Table 1-5 Diffraction-Limited MTF Response

K	T	K	T
0.02	0.97	0.52	0.37
0.04	0.95	0.54	0.35
0.06	0.92	0.56	0.33
0.08	0.90	0.58	0.31
0.10	0.87	0.60	0.28
0.12	0.85	0.62	0.26
0.14	0.82	0.64	0.24
0.16	0.80	0.66	0.23
0.18	0.77	0.68	0.21
0.20	0.75	0.70	0.19
0.22	0.72	0.72	0.17
0.24	0.70	0.74	0.15
0.26	0.67	0.76	0.14
0.28	0.65	0.78	0.12
0.30	0.62	0.80	0.10
0.32	0.60	0.82	0.09
0.34	0.58	0.84	0.07
0.36	0.55	0.86	0.06
0.38	0.53	0.88	0.05
0.40	0.50	0.90	0.04
0.42	0.48	0.92	0.03
0.44	0.46	0.94	0.02
0.46	0.44	0.96	0.01
0.48	0.41	0.98	0.00
0.50	0.39		

LASER OPTICS

A strictly Gaussian beam in its fundamental transverse mode has a beam profile given by (see O'Shea, 1985; Siegman, 1971)

$$I = I_0 e^{-2R^2/W^2}$$

where I_0 is the intensity at the center of the beam, R is a distance measured from the beam center, and W is the beam radius where the intensity is reduced to $1./e^2 I_0$. This beam profile is shown in Fig. 1-9.

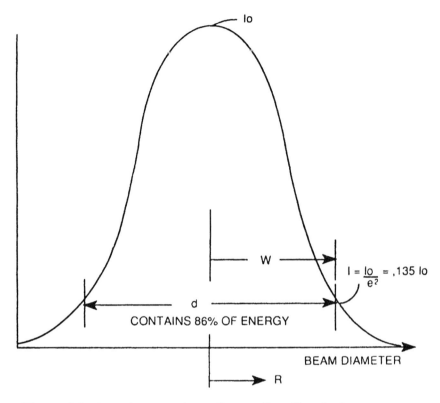

Figure 1-9 Intensity versus beam diameter for a Gaussian beam.

Most gas laser systems have such a Gaussian beam profile. When the laser manufacturer lists the beam diameter as d ($d=2W$) it corresponds to the $1/e^2$ intensity value (0.135). Within this diameter is 86% of the power. Within a diameter of $1.5d$ there is 99% of the power. It is thus common practice to design laser optical systems (focusing lenses, beam expanders, etc.) to accommodate this $1.5d$ (for very high power systems $2d$ is often used; it then contains 99.97% of the total power).

Within close proximity to the laser is the beam waist. Beyond this waist, the beam expands (hyperbolic). The full angle beam divergence at a distance from this waist is given by (Kogelnik, 1966)

$$\frac{1.27\lambda}{d}$$

Table 1-6 Some Common Laser Lines

Laser	Strongest lines (microns)
ArF Excimer	0.1935
KrF Excimer	0.248
XeCl Excimer	0.308
XeF Excimer	0.351
Helium neon	0.6328
Helium cadmium	0.4416
Carbon dioxide	10.59
Argon ion	0.4880, 0.5145 (lines from 0.45–0.53)
Krypton ion	0.6471, 0.6764 (lines from 0.46–0.68)
Ruby	0.6943
Neodymium	1.06

REFERENCES

American National Standards Institute (1980). Definitions, Methods of testing, and Specifications for Appearance Imperfections of Optical Elements and Assemblies, PH3.617, New York.

Betensky, E. (1980). *Photographic Lenses, Applied Optics and Optical Engineering,* Vol. 8 (R. Shannon and J. Wyant, eds.), Academic Press, New York.

Born, M. and Wolf, E. (1965). *Principles of Optics,* Pergamon Press, New York.

Brixner, B. (1978). The merit function in lens design, *Appl. Opt., 17:*715.

Busby, E. S. (1972). Variable light transmitting filter for cameras, U.S. Patent #3700314.

Conrary, A. E. (1957). *Applied Optics and Optical Design,* Dover, New York.

Cox, A. (1964). *A System of Optical Design,* Focal Press, London.

Feder, D. (1957). *Automatic lens design methods, JOSA,* 47:902.

Fischer R. E., ed. (1978). *Computer Aided Optical Design,* SPIE, Bellingham, WA.

Fischer, R. E. ed. (1980). *International Lens Design Conference,* SPIE, Bellingham, WA.

Ginsberg, R. H. (1981). An outline of tolerancing, *Optical Engineering, 20:*175.

Grey, D. (1963). Aberration theories for semiautomatic lens design by electronic computers, *JOSA, 53:*672.

Grey, D. (1970). Tolerance sensitivity and optimization, *Applied Optics, 9:*523.
Grey, D. (1978). The inclusion of tolerance sensitivities in the merit function for lens optimization, *SPIE, 147:*63.
Herman, R. M. (1985). Diffraction and focusing of Gausian beams, *Appl. Opt., 24:*1346.
Hopkins, H. H. and Tiziani, H. J. (1966). A theoretical and experimental study of lens centering errors, *Brit. J. Appl. Physics, 17:*33.
Hopkins, R., McCarthy, and Walters, R. M. (1955). *Automatic correction of third order aberrations, JOSA, 45:*365.
Howard, J. W. and Abel, I. R. (1982). Narcissus; reflections on retroreflections in thermal imaging systems, *Applied Optics, 21:*3393.
SO Standard 10110 (1993). Distributed by the National Association of Photographic Manufacturers, Harrison, New York.
Institute of Optics (1967). Lens design with large computers. Proceedings of the Conference, Rochester, NY.
International Lens Design Conference (1985) at Cherry Hill, NJ. Technical Digest, SPIE, Bellingham WA.
Kimmel, R., and Parks, R. E. (1995). ISO 10110 Optics and Optical Instruments, American National Standards Institute, New York, NY.
Kingslake, R. (1978). *Lens Design Fundamentals,* Academic Press, New York.
Kingslake, R. (1983). *Optical System Design,* Academic Press, New York.
Kingslake, R. (1989). *History of the Photographic Lens,* Academic Press, New York.
Koch, D. G. (1978). A statistical approach to lens tolerancing, *SPIE, 147:*71.
Kodak. Optical formulas and their application, Kodak Publication AA-26.
Kogelnik, H. and Li, T. (1966). Laser beams and resonators, *Proc. IEEE, 54:*1312; *Applied Optics* 5:1550.
Lavi, A. and Vogl, T. P., eds. (1966). *Recent Advances in Optimization Techniques,* John Wiley, New York.
Lawson, L. L. and Hanson, R. J. (1974). *Solving Least Squares Problems,* Prentice Hall, Englewood Cliffs, NJ.
Malacara (1978). *Optical Shop Testing,* John Wiley, New York.
Meiron, J. (1959). Automatic lens design by the least squares method, *JOSA, 49:*293.
Military Standard. Preparation of drawings for optical elements, MIL-STD-34.
Military Standardization Handbook, MIL HBK-141 (1962). U.S. Govt. Printing Office, Washington, DC 20402.

Norland Products Catalog (1999). New Brunswick, NJ 08902.

O'Shea, D. C. (1977). *Introduction to Lasers and Their Application,* Addison-Wesley, New York.

O'Shea, D. C. (1985). *Elements of Modern Optical Design,* Wiley Interscience, New York.

O'Shea, D. C. and Thompson, B. J., eds. (1988). *Selected Papers on Optical Mechanical Design,* SPIE, Bellingham, WA.

Palmer, J. M. (1971). *Lens Aberration Data,* American Elsevier Publishing.

Reitmayer, F. and Schroder, H. (1975). Effect of temperature gradients on the wave aberration in athermal optical systems, *Applied Optics 14:*716.

Rimmer, M. P. (1978). A tolerancing proceedure based on modulation transfer function, *SPIE, 147:*66.

Rosen, S. and Eldert, C. (1954). Least squares method for optical correction, *JOSA, 44:*250.

Siegman, A. E. (1971). *Introduction to Lasers and Masers,* McGraw-Hill, New York.

Smith, W. (1966). *Modern Optical Engineering,* McGraw-Hill, New York.

Starke and Wise (1980). MTF based optical sensitivy and tolerancing programs, *Applied Optics, 19:*1768.

Summers Laboratories Catalog (1991). P.O. Box 162, Fort Washington, PA 19034.

Tamagawa, Y. and Wakabayashi, S. (1994). System with athermal chart, *Applied Optics, 33:*8009.

Tamagawa, Y. and Tajime, T. (1996). Expansion of an athermal chart into a multilens system, *Optical Engineering, 35:*3001.

Tatian, B. (1984). Fitting refractive index data with the scllmeier dispersion formula, *Applied Optics, 23:*4477.

Tuchin, G. D. (1971). Summing of optical systems aberrations caused by decentering, *Sov. J. Optical Tech., 38:*546.

Unvala, H. A. (1966). The orthonormalization of aberrations, AD-640395, available from NTIS, Springfield, VA.

Wang, J. (1972). Tolerance conditions for aberrations, *JOSA, 62:*598.

Welford, W.T. (1986). *Aberrations of Optical Systems,* Adam Hilger, Bristol.

Willey, R. R. (1990). Antireflection coating for high index cemented doublets, *Applied Optics, 29:*4540.

2
The Achromatic Doublet

Consider two thin lenses with a real and distant object with a combined effective focal length of F. Let F_a be the focal length of the 'a' lens (facing the long conjugate) with a V value of V_a, and likewise for the 'b' lens. Then (Kingslake, p. 80)

$$F_a = \frac{(V_a - V_b)F}{V_a} \quad \text{and} \quad F_b = \frac{(F_b - F_a)F}{V_b}$$

Using thin-lens G sum formulae (Smith, 1966, p. 281; Ingalls, 1953, p. 208) to obtain a lens system with zero third-order spherical aberration and coma, we obtain the following algorithm for a thin lens achromatic doublet.

Select materials: V_a, N_a, V_b, N_b, and system focal length F.
Determine F_a and F_b from the above equations.

$$C_a = \frac{1}{F_a(N_a - 1)} \quad C_b = \frac{1}{F_b(N_b - 1)}$$

$$H = (G8b)(C_b)^2 - (G8a)C_a^2 - \frac{(G7b)C_b}{F}$$

$$I = \frac{(G5a)C_a}{4} \quad K = \frac{(G5b)C_b}{4}$$

$$A = (G1a)C_a^3 + (G1b)C_b^3 - \frac{(G3b)C_b^2}{F} + \frac{(G6b)C_b}{F^2}$$

$$B = -(G2a)C_a^2 \quad E = (G4a)C_a \quad J = (G4b)C_b$$

45

$$D = (G2b)C_b^2 - \frac{(G5b)C_b}{F} \qquad P = \frac{A + H(JH/K - D)}{K}$$

$$Q = B + \frac{I(2JH/K - D)}{K} \qquad R = E + J(I/K)^2$$

$$ROOT = Q^2 - 4PR \qquad \text{(Check for negative root)}$$

$$C1 = \frac{-Q + \sqrt{ROOT}}{2R} \qquad \text{(King 1993)}$$

$$C4 = -\frac{(H + IC1)}{K} \qquad C2 = C1 - C_a \qquad C3 = Cb + C4$$

In the above equations, $G1a$ is the $G1$ sum for the 'a' lens, $G8b$ is the $G8$ sum for the 'b' lens, etc. (see Smith, 1966). That is, $G8 = N(N-1)/2$, etc. C1, C2, C3, and C4 are the surface curvatures.

This algorithm has been programed on the author's computer to arrive rapidly at a thin lens third-order solution as a start for lens optimization. (See the program Design an Achromat in Appendix E.) A similar program is discussed by Reidl (1981). Several examples are now given for a focal length of 10.

Table 2-1 Thin Lens Third-Order Solution for an Achromat

Na	Va	Nb	Vb	C1	C2	C3	C4
4.0031	701.4	2.4054	34.20	0.1005	0.0655	−0.0378	−0.0341
1.5187	63.96	1.6522	33.60	0.1646	−0.2415	−0.2404	−0.0707
1.5712	55.85	1.6241	36.11	0.1573	−0.3380	−0.3352	−0.0421
1.5749	57.27	1.6942	30.95	0.1566	−0.2219	−0.2217	−0.0523
1.4610	8.648	1.4980	6.195	0.1726	−0.5921	−0.5835	−0.0764
3.4313	420.2	4.0301	188.4	0.1046	0.0300	0.0241	0.0509
2.2524	233.1	1.3572	29.53	0.1123	0.0209	−0.0500	−0.0094

The first case is germanium and zinc selenide, 8 to 14μ region. Then follow three combinations for the visual region; BK7 and SF2 glass, BaK4 and F2, BAK1 and SF8. Next is a UV doublet with CaF_2 and quartz (silica). The last two cases are for the infrared region 3.2 to

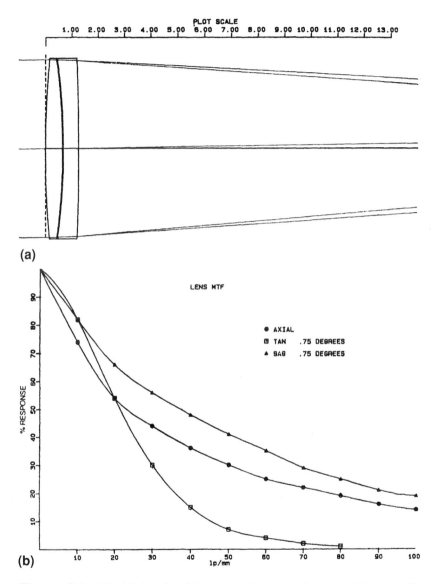

Figure 2-1 (a) 48 in. focal length achromat telescope objective f/8. (b) Lens MTF.

4.2 µ; a silicon–germanium combination, followed by IRTRAN 2–IRTRAN 1 doublet.

The second case was scaled to a focal length of 48 for an f/8 telescope objective with a field of view of 1.5 degrees. It was optimized and is shown as Fig. 2-1. In Table 2.2 is the data for this lens.

Table 2-2 48 inch EFL, f/8 Telescope Objective

Radius	Thickness	Diameter	Material
29.2097	0.655	6.10	BK7
–19.9047	0.032	6.10	AIR
–19.8979	0.578	6.00	SF2
–66.3937	47.381	6.10	AIR

The distance from the first lens surface to the image is 48.646.

Figure 2-2 shows a cemented achromat, f/6 of 20 inch focal length. Field of view is 1.5 degrees.

Radius	Thickness	Diameter	Material
12.8018	0.434	3.41	BAK1
–9.0623	0.321	3.41	SF8
–37.6553	19.631	3.41	AIR

The distance from the first lens surface to the image is 20.386.

In both of these designs, the entrance pupil is in contact with the first lens surface. As is typical with such designs, the off-axis sagittal MTF is better than tangential. Due to the small field of view and large f#, the main aberrations here are secondary color and field curvature. Longitudinal secondary color is approximately focal length/2000.

The radius of the Petzval surface (for a single thin lens) = $-NF$. However, in the above doublet case, $Rp = -1.45F$. Due to the presence of astigmatism, the best image surface radius is substantially shorter than this, approximately 0.48 focal length.

These same equations may be used to obtain a lens of negative

The Achromatic Doublet

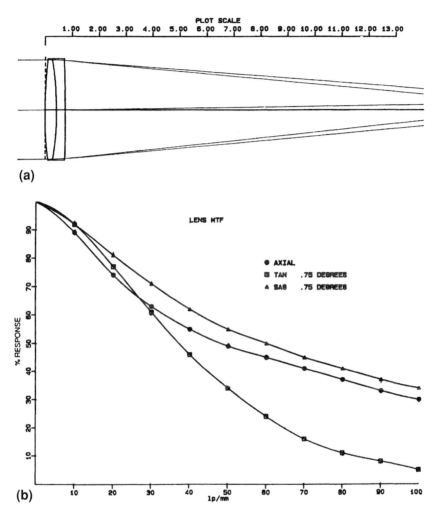

Figure 2-2 (a) Cemented achromat $f/6$. (b) Lens MTF.

focal length, the so called Barlow lens (Ingalls, 1953). For a focal length achromat of −10, simply change the signs of the curvatures as given in Table 2-1.

As discussed earlier, in order to reduce secondary color, a material off the glass line must be selected. For three thin lenses in contact, the value

$$\sum \frac{\varphi_i P_i}{V_i}$$

must be zero. Where φ_i is the lens element power (Knetsch, 1970),

$$V_i = \frac{N_e - 1}{N_f - N_c} \quad \text{and} \quad P_i = \frac{N_f - N_e}{N_f - N_c}$$

In Fig. 2-3 is shown a 10 inch focal length cemented achromat. It is f/5 and covers a 3 degree field of view. Data for this lens is

Surf	Radius	Thickness	Material	Diameter
0	0.0000	1.0000E+10		
1	Stop	0.0000		2.000
2	6.5042	0.3500	BAK1	2.000
3	−4.9573	0.2000	SF1	2.000
4	−17.1987	9.7266		2.000
5	0.0000	0.0000		0.526

Following (Fig. 2-4) is a cemented, three element lens with greatly reduced secondary color.

Radius	Thickness	Diameter	Material
7.3324	0.417	3.44	SSKN5
4.7929	0.655	3.44	FPL53
−18.2225	0.600	3.44	BAK1
−44.8800	18.846	3.44	AIR

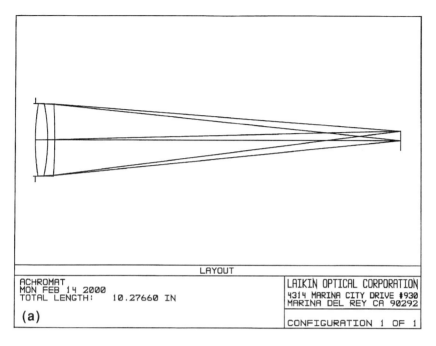

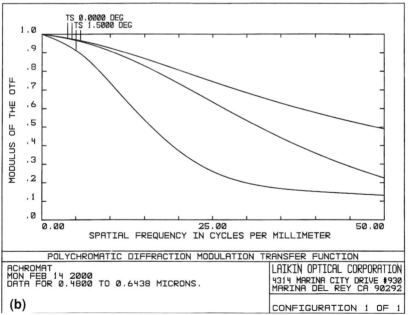

Figure 2-3 (a) 10 in. focal length cemented achromat $f/5$. (b) Lens MTF.

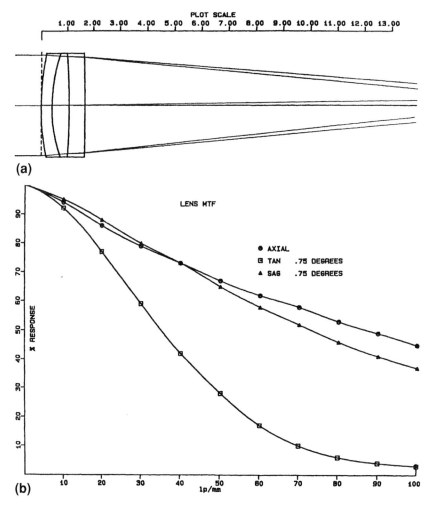

Figure 2-4 (a) Cemented apochromat $f/6$. (b) Lens MTF.

The distance from the first lens surface to the image is 20.519. The entrance pupil is in contact with the first lens surface. Like the above achromat (Fig. 2-2), it is f/6, 20 inch focal length and has a field of view of 1.5 degrees. It is shown in figure 2-4a, MTF in Fig. 2-4b. The longitudinal secondary color is focal length/7400. This greatly improves the axial MTF. However, off axis is limited by astigmatism, which is about the same as for the above achromat. Strangely enough, an achromatic lens (actually a "pseudo" achromat) may be made with only one type of glass. Consider two thin lenses F_1 and F_2 separated by a distance D. The back focal length, BFL, is given by

$$BFL = \frac{(F_1 - D)}{F_1 + F_2 - D}$$

Differenting to determine $\partial BFL/\partial \lambda$ and setting this to zero,

$$D = F_1 \pm \sqrt{-F_1 F_2}$$

For example, if $N = 1.5$, $F_1 = 20$, and $F_2 = -10$, then $D = 5.8578$, EFL = -48.284, and BFL = -34.142. Note that this is not a true achromat since two wavelengths are not united at a common focus. The BFL change with wavelength is now a minimum, so this is only of value over a limited wavelength region. Also the negative EFL and BFL is a problem. See also Malacara (1994).

REFERENCES

Hariharan, P. (1997). Apochromatic lens combinations, *Optics and Laser Technology*, 4:217.

Hastings, C.S. (1889). Telescope objective, U.S. Patent #415040.

Hopkins, R.E. (1995). Automatic design of telescope objectives, *JOSA*, 45:992.

Ingalls, A.G., ed. (1953). *Amateur Telescope Making*, Book 3, Scientific American.

King, S. (1993). Personal correspondence. I would like to thank him for pointing out a subtle error in a previous edition.

Kingslake, R. (1978). *Lens Design Fundamentals,* Academic Press, New York.

Knetsch, G.(1970). Three lens objective with good correction of the secondary spectrum, U.S. Patent #3536379.

Kutsenko, N.I. (1975). The calculation of thin three lens cemented components, *Sov. J. of Opt. Tech., 45:*82.

Lessing, N. W. (1957). Selection of optical glasses in apochromats, *JOSA, 47:*955.

Malacara, D. and Malacara, Z. (1994). Achromatic aberration correction with only one glass, *SPIE, 2263:*81.

Reidl, M. (1981). The thin achromat, *Electro-Optical Systems Design,* Sept. 1981, p. 49.

Smith, W. (1966). *Modern Optical Engineering,* McGraw-Hill, New York.

Szulc, A. (1996). Improved solution for the cemented doublet, *Applied Optics, 35:*3548.

Uberhagen, F. (1970). Doublet which is partially corrected spherically, U.S. Patent #3511558.

3
The Air-Spaced Triplet

The air-spaced triplet is sometimes referred to as a Cooke triplet. It was developed by Harold Dennis Taylor in 1894 (he worked for a company in York, England, T. Cooke and Sons). There are enough degrees of freedom to design an anastigmat lens. Referring to Fig. 3-1, for a series of three thin lenses (note that although the lenses are thin, the air spaces separating them are appreciable), as a start we control system power and set Petzval sum, longitudinal chromatic, and lateral color to zero. Let us also assume that the material for the first and third lenses is the same, that the stop is at the second lens, and that the object is distant.

$$F_a = \frac{1}{P_a}$$

where P_a is the power of the first lens, etc.

$$TR = \frac{T_2}{T_1}$$

$$T_3 = \frac{\left[(F_a - T_1)F_b - (F_a + F_b - T_1)T_2\right]F_c}{(F_a - T_1)F_b + (F_a + F_b - T_1)(F_c - T_2)} \quad (1)$$

$$X = T_3 \, PTR$$

where P is the power of the lens assembly.

$$P_b = \frac{1}{F_b} = -P_c N_b \frac{X+1}{N_a} \quad (2)$$

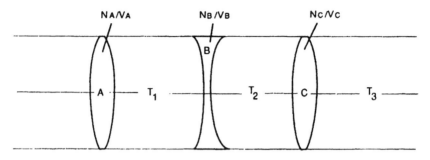

Figure 3-1 Triplet lens.

$$P_a = XP_c \tag{3}$$

$$P = P_a + P_b + P_c - T_1 P_a (P_b + P_c) - T_2 P_c (P_a + P_b) + T_1 T_2 P_a P_b P_c \tag{4}$$

$$P_c = \frac{1/X - \sqrt{\left[(X/TR^2 + 1)V_b N_a / (V_b N_a) / (V_a X N_b (X+1))\right]}}{T_1} \tag{5}$$

Our algorithm becomes an iterative technique. Let E be a very small number (the error of our iteration). Given values of N_a, V_a, N_b, V_b, TR, and system power P, as a start, let $T_1 = 0.1/P$, $X = 0.8$. Then $T_2 = TR T_1$.

```
        DO 7 J= 1,10
        from Eq. (5) calculate P_c;
        from Eq. (2) calculate P_b;
        from Eq. (3) calculate P_a;
        from Eq. (1) calculate T_3;
        from Eq.(4) calculate P;
        X1 = T_3 P TR
        IF (ABS(X1–X) –E) 8,8,6
6       X = X1
7       CONTINUE
8       S = P/Φ
```

The Air-Spaced Triplet

$$P_a = P_a/S$$
$$P_b = P_b/S$$
$$P_c = P_c/S$$
$$T_1 = T_1 \, S$$
$$T_3 = T_3 \, S$$
$$T_2 = T_1 \, TR$$

This algorithm was programed on the author's computer and is routinely used for triplet starting solution. See the program Design a Triplet in Appendix E.

Table 3-1 gives six such solutions, F=100, and assumes that $T_2 = T_1$ ($TR = 1$).

Table 3-1 Thin Lens Triplet Starting Solutions

P_a	P_b	P_c	T_1	T_3	Material
0.02244	−0.04804	0.02556	12.203	87.797	SK16, F2
0.03347	−0.06525	0.03480	3.799	96.201	SK16, LLF1
0.02088	−0.04913	0.02514	16.938	83.062	Plexiglas, polystyrene
0.03505	−0.07391	0.03703	5.327	94.673	CaF_2, quartz
0.01873	−0.05124	0.02490	24.788	75.212	Silicon, germanium
0.00938	−0.03296	0.04315	78.262	21.737	IRTRAN2, IRTRAN3

The first three solutions are for the visual region, the next is for the UV region, and the last two are for the IR region of 3.2 to 4.2 microns. This algorithm does not converge for materials like germanium and zinc selenide. See Vogel (1968). This table illustrates the advantage of using this third-order technique as a preliminary design tool. Note that by reducing the refractive index of the middle element of the second solution as compared to the first, we obtain a more compact system with a longer BFL.

As a start in optimization, it is expedient to hold the powers and spacing of the elements while bending the lenses to solve for spherical, coma, and astigmatism. Next, thicken the elements and vary all the parameters.

Optimizing the first solution yields the design shown in Fig. 3-2.

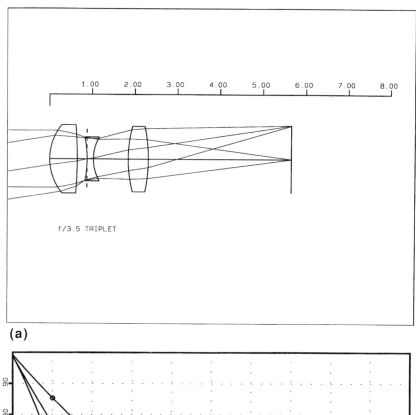

(a)

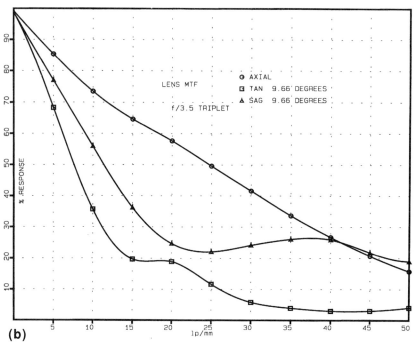

(b)

Figure 3-2 (a) f/3.5 triplet. (b) Lens MTF:

It is f/3.5 and was designed to project 24 × 36 mm film. EFL = 5. The stop is in contact with surface 3.

Table 3-2 f/3.5, EFL = 5 triplet

	Radius	Thickness	Material	Diameter
	Entrance pupil	−0.8771		1.428
1	1.42100	0.6499	SK16	1.720
2	−10.90452	0.2300	Air	1.720
3	−2.93835	0.1500	F2	1.100
4	1.19190	0.8113	Air	1.100
5	3.48672	0.4756	SK16	1.650
6	−4.26794	3.3393	Air	1.650

The distance from the front lens vertex to the image is 5.656. In order to reduce the Petzval sum, designs of this type have glasses such that the front and rear elements are made from a high index crown, while the center element is a low index flint (Sharma, 1982). Distortion = 0.67%.

After a few small changes in Vogel's prescription, and then an optimization, we get the solution for an infrared lens (8 to 14 μ as shown in Fig. 3-3.

Table 3-3 Infrared Lens, 8 to 14 μ

Radius	Thickness	Diameter	Material
10.1950	0.590	5.16	Germanium
13.6008	5.251	5.00	Air
−10.3250	0.590	3.48	Zinc selenide
−11.4579	4.735	3.60	Air
2.5751	0.392	2.34	Germanium
2.4867	1.307	2.00	Air

The entrance pupil is in contact with the first lens surface. It is f/2 and has a focal length of 10 and a FOV of 8 deg. The distance of the first lens surface to the image is 12.865. Distortion = 0.2 %. It is interesting to note that the rear lens is actually a negative lens and so is act-

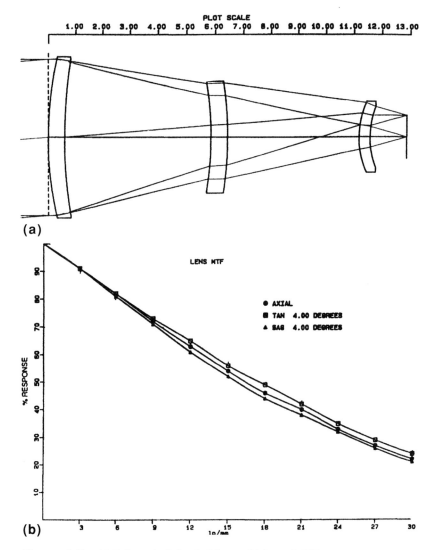

Figure 3-3 (a) Infrared triplet 8–14 μm. (b) Lens MTF:

ing as a field flattener. Considering the long wavelength, this lens is nearly diffraction limited.

Table 3-4 EFL = 4, Silicon-Germanium Triplet Lens

Radius	Thickness	Diameter	Material
	−12.069	1.00	Entrance pupil
4.8328	0.222	2.56	Silicon
10.8198	0.898	2.56	Air
−11.1003	0.100	1.69	Germanium
25.3859	1.192	1.69	Air
12.1820	0.152	1.60	Silicon
−16.3424	2.774	1.60	Air

Figure 3-4 is the result of an optimization of the fifth solution (silicon-germanium, 3.2 to 4.2 μ) EFL = 4. Following is the lens prescription.

This lens is f/4 with a field of view of 7 degrees. First lens surface to image, 5.338. Note from the schematic that the aperture stop is behind the lens. In some applications, this can be an advantage, but it does cause the front element to be large. Distortion is less than 0.2% and is pincushion.

In Fig. 3-5 is shown a 50 inch focal length f/8 triplet, which covers a 2.25 × 2.25 (3.18 diagonal) film format. It was designed to be used for photography by amateur astronomers. Distortion is negligible.

Note that, unlike the triplet of Fig. 3-2, the aperture stop is in front of the lens, and the center element is very thick. To make a practical device, this element should be thinned down a little.

Table 3-5 50 inch EFL Triplet

Radius	Thickness	Diameter	Material
	0.321	6.25	Aperture stop
15.9265	0.357	6.39	SK16
56.8199	3.389	6.39	Air
−19.8503	1.332	6.00	LF5
20.9083	1.902	6.08	Air
87.7652	0.620	6.50	SK16
−16.1177	47.211	6.50	Air

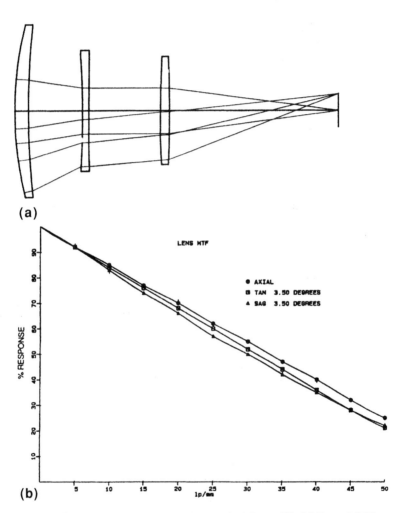

Figure 3-4 (a) 4-in. infrared triplet 3.2–4.2 μm $f/4$. (b) Lens MTF.

The Air-Spaced Triplet

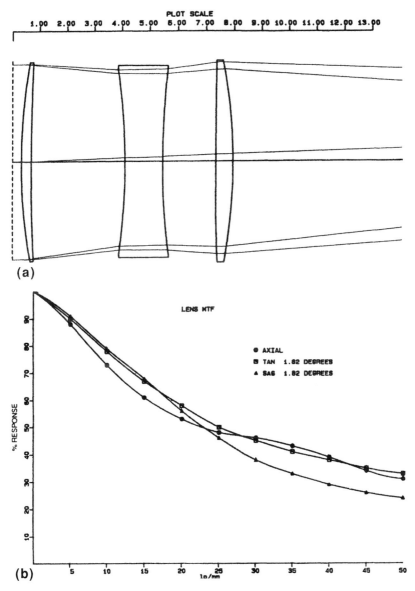

Figure 3-5 (a) 50-in. f8 triplet. (b) Lens MTF.

The distance of the first lens surface to the image is 54.810. The limiting aberration is longitudinal secondary color. This is nearly 0.7 mm.

REFERENCES

Ackroyd, M. D. (1968). Wide angle triplets, U.S. Patent #3418040.

Ackroyd, M. D. (1969). Triplet type projection lens, U.S. Patent #3443864.

Arai, Y. (1980). Achromatic objective lens, U.S. Patent #4190324.

Conrad, L. H. (1972). Three element microphotographic objective lens, U.S. Patent #3640606 and #3640607.

Eckhardt, S. K. (1997). Fixed focus, triplet projection lens for overhead projectors, U.S. Patent #5,596,455.

Kallo, P. and Kovacs, G. (1993). Petzvzl sum in triplet design, *Optical Engineering, 32:*2505.

Kingslake, R. (1968). Triplet covering a wide field, U.S. Patent #3418039.

Kobayashi, K. (1969). Ultra-achromatic fluorite silica triplet, U.S. Patent #3486805.

Sharma, K. D. (1982). Utility of low index high dispersion glasses for Cook triplet design, *Applied Optics, 21:*1320.

Sharma, K. D. and Gopal, S. V. (1982). Significance of selection of Petzval curvature in triplet design, *Applied Optics, 21:*4439.

Stephens, R. E. (1948). The design of triplet anastigmat lenses, *JOSA, 38:*1032.

Tronnier, E. (1965). Three lens photographic objective, U.S. Patent #3176582.

Vogel, T. (1968). Infrared optical system, U.S. Patent #3363962.

4
Triplet Modifications

To reduce f#, the front element is often split into two positive lenses. Another modification involves the splitting of the rear element into a cemented negative–positive lens (Tessar). Still another modification splits both front and rear elements into cemented negative–positive assemblies (Heliar). A modification of this last form is shown in Fig. 4-1. It is f/5, 20 degree FOV, and has a focal length of 10. Visual region.

Table 4-1 Heliar, f/5

	Radius	Thickness	Material	Diameter
0	0.00000	–2.8782	Entrance pupil	
1	4.13088	0.8890	SK16	2.980
2	–4.03166	0.2999	LLF6	2.980
3	13.80239	0.8820	Air	2.360
4	–3.16214	0.2000	LLF6	1.780
5	3.07900	0.3110	Air	1.660
6	Stop	0.0533	Air	1.617
7	14.13740	0.2500	LLF6	2.100
8	2.33444	0.7474	SK16	2.100
9	–3.25924	8.5118	Air	2.100

The distance from the front lens surface to the image is 12.144. Distortion is less than 0.1%.

Figure 4-2 shows a four inch focal length, f/4.5 Tessar lens. It covers a field of 3.0 diameter.

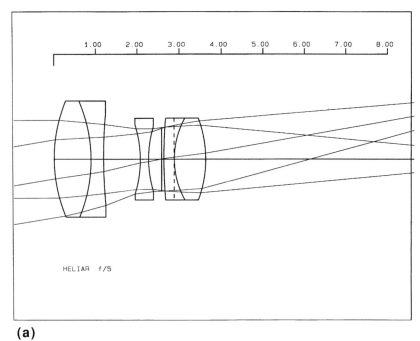

(a)

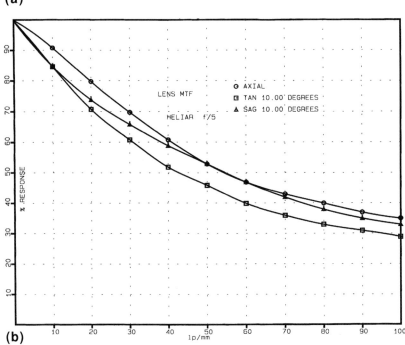

(b)

Figure 4-1 *f*/5, 20° FOV. (b) Lens MTF.

Triplet Modifications

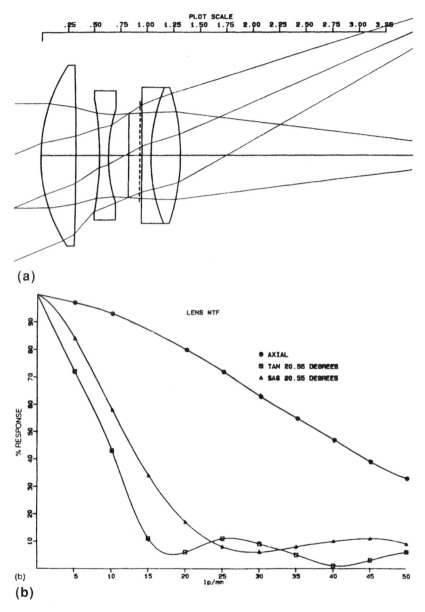

Figure 4-2 (a) Tessar lens 4-in. *f*/4.5. (b) Lens MTF.

Table 4-2 f/4.5 Tessar

Radius	Thickness	Diameter	Material
	−0.930	0.89	Entrance pupil
1.3147	0.329	1.54	SK15
−21.0190	0.223	1.54	Air
−2.4330	0.081	0.98	F2
1.1718	0.194	0.81	Air
Stop	0.125	0.72	Air
−31.8418	0.082	0.90	K10
1.4005	0.278	1.16	SK15
−1.6991	3.304	1.16	Air

The distance from the first lens surface to the image is 4.617. Distortion = 0.4%.

In the above examples, note the use of high index crown for front and rear elements, and low index flint for the center element (to reduce Petzval sum as with a triplet). Tessar type lenses are often used as enlarging lenses. This design is similar to the Velesek (1975) patent (Chap. 21 references). In both cases, the rear cemented assembly has a small V_d difference and a strong curvature between the elements.

In Fig. 4-3 is shown a lens for use with a slide projector for 35

Table 4-3 Slide Projector Lens

Radius	Thickness	Diameter	Material
	−2.990	1.43	Entrance pupil
10.2481	0.306	2.59	SK4
−7.1216	0.022	2.59	Air
1.6698	0.576	2.13	SK4
4.0784	0.410	1.72	Air
−2.9088	0.090	1.35	SF8
1.2293	0.632	1.12	Air
Stop	0.023	1.05	Air
5.4592	0.256	1.22	SK4
−1.5109	2.879	1.22	Air

The distance from the first lens surface to the film is 5.194. Distortion = 0.1%. There is 10% vignetting at the edge of the field.

Triplet Modifications

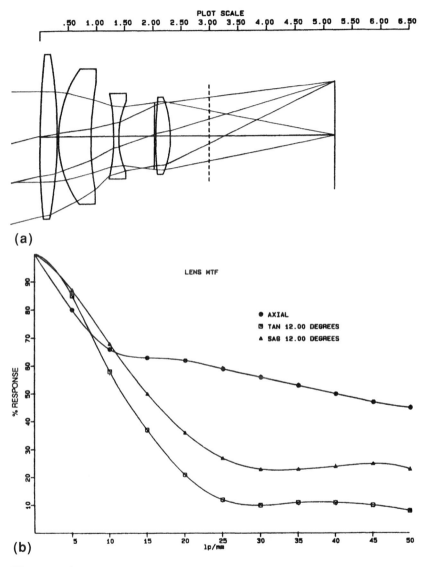

Figure 4-3 (a) Slide projector lens. (b) Lens MTF.

mm SLR film. Referring to Appendix A, we see that this has a diagonal of 1.703. Since this lens has a focal length of 4, the field of view then is 24 degrees. The lens is f/2.8

If a target of 33 lp/mm were projected with this lens, the screen image of these lines would subtend one minute of arc for an observer at the projection lens. Realizing that in a darkened room in which one would project slides, one cannot resolve one minute of arc (probably 2 to 3 minutes of arc would be more realistic), this lens performance is excellent.

REFERENCES

1. Cook, G.H. (1950). Highly corrected three component objectives, U.S. Patent #2502508.
2. Doi, Y. (1981). Rear stop lens system, U.S. Patent #4298252.
3. Edwards, G.(1972). Four component objective, U.S. Patent #3649104.
4. Eggert, J.(1965). Objective lens consisting of four lens units, US. Patent # 3212400.
5. Guenther, R. E. (1970). Four element photographic objective, U.S. Patent #3517987.
6. Hopkins, R. E. (1965). Optical lens system design, AD 626844 Defense Documentation Center).
7. Mihara, S. (1984). Compact camera lens system with a short overall length, U.S. Patent #4443069.
8. Sharma, K. D. (1979). Design of new 5 element Cooke triplet derivative, *Applied Optics, 18:*3933.
9. Sharma, K. D. (1980). Four element lens systems of the Cooke triplet family, *Applied Optics, 19:*698.
10. Tateoka, M. (1983). Projection lens, U.S. Patent #4370032.
11. Tronnier, A. W. (1937). Unsymmetrical photographic objective, U.S. Patent #2084714.

5
Petzval Lenses

This class of lenses consists of two positive members separated by a large air space of perhaps half of the focal length of the lens. The system exhibits a large negative Petzval sum, has an inward curving field, and so is only usable for small fields of view. It is of economical construction and is capable of a low f#.

Figure 5-1 shows a basic f/1.4 projection lens for 16 mm motion picture film. It has a 14 degree FOV and a focal length of 2.

Table 5-1   f/1.4 Projection Lens

	Radius	Thickness	Material	Diameter
0	0.00000	1.00000E+07	Air	
1	1.39649	0.3597	LAKN12	1.560
2	−3.24586	0.0445	Air	1.560
3	−2.36948	0.3983	SF8	1.480
4	2.73116	0.7215	Air	1.280
5	1.09027	0.6167	LAKN12	1.240
6	−2.12417	0.1691	Air	1.240
7	−0.88883	0.2000	SF4	0.760
8	38.90132	0.3963	Air	0.700

The distance from the front lens surface to the image is 2.906. Distortion = 0.4%. The entrance pupil is −0.258 from the first lens surface.

In Fig. 5-2 is shown a Petzval lens in which the spacing between the last two elements has been substantially increased. The focal

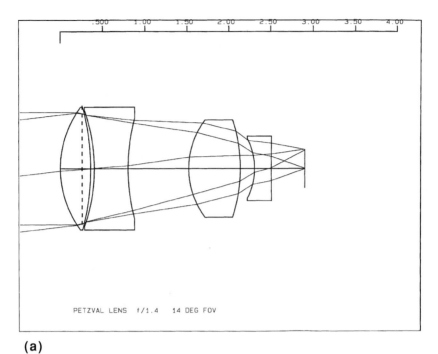

(a)

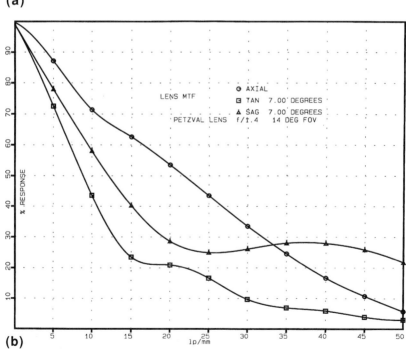

(b)

Figure 5-1 (a) Petzval lens $f/1.4$. (b) Lens MTF.

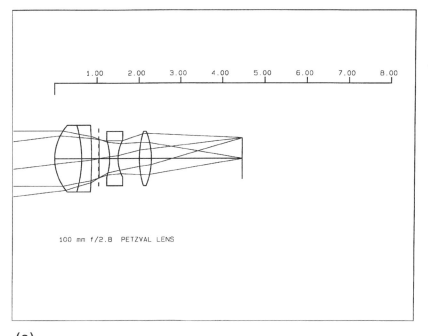

(a)

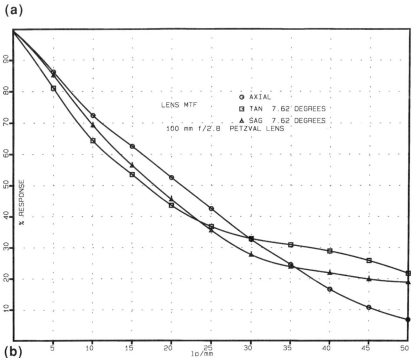

(b)

Figure 5-2 (a) 100 mm *f*/2.8 Petzval lens. (b) Lens MTF.

length is 100 mm and is f/2.8; The field of view is 15.24 degrees. It may be used as either a camera lens or a projection lens for 35 mm cinematography.

Table 5-2   100 mm f/2.8 Lens

	Radius	Thickness	Material	Diameter
0	0.00000	1.00000E+07	Air	
1	1.37404	0.6350	SK4	1.700
2	−3.16813	0.2354	SF1	1.700
3	−13.01628	0.1825	Air	1.700
4	Stop	0.2500	Air	0.984
5	−1.62596	0.1968	F5	0.940
6	1.04471	0.5113	Air	0.940
7	2.47184	0.2924	SK4	1.380
8	−1.95684	2.1622	Air	1.380

The distance from the first lens surface to the image is 4.465. Distortion = 0.39%.

REFERENCES

1. Rogers, P. J. (1980). Modified Petzval lens, U.S. Patent #4232943.
2. Shade, W. (1951). Objectives of the Petzval type with a high index collective lens, U.S. Patent #2541484.
3. Smith, W. J. (1966). Objective of the Petzval type with field flattener, U.S. Patent #3255664.
4. Werfeli, A. (1956). Photographic and projection objective of the Petzval type, U.S. Patent #2744445.

6
Double Gauss Lenses

This type of lens was derived from a telescope objective originally designed by K. F. Gauss. It is one of the basic forms of anastigmatic photographic lens and consists of a nearly symmetric arrangement of elements about a central stop. Surrounding this stop are two achromats with the flint element facing the stop (Brandt, 1956) such that its surface is concave toward the stop. It can be well corrected over a large aperture and a moderately large field of view.

Apertures are generally f/2.8 or less, with fields of view of at least 30 degrees, and a BFL of 0.5 to 0.9 EFL. It is the basic lens supplied (50 mm focal length) with 35 mm SLR cameras. To reduce the f#, additional elements are generally added in the front and rear.

In Fig. 6-1 is shown an f/2.5 double Gauss lens with a 40 deg FOV and an EFL of 5.

Figure 6-2 shows a 50 mm f/1.8 SLR camera lens. It is the basic lens supplied with 35 mm single lens reflex (SLR) cameras. The large back focal length required is to clear the shutter and reflex mirror mechanism in the camera. A great deal of effort has been spent in developing lenses for these cameras, which are compact, are light in weight, and have good performance. See, for example, Wakamiya (1984).

At full aperture there is about 40% vignetting at the corner of the field and 20% vignetting at 16 degrees off axis. There is no vignetting at f/4. Focusing is generally done at full aperture in the center of the field. Then, at the time of exposure, the lens is stopped down. So it is important to have relatively low spherical aberration to prevent a focal shift as the lens is stopped down. In Table 6-2 is the lens data for Fig. 6-2.

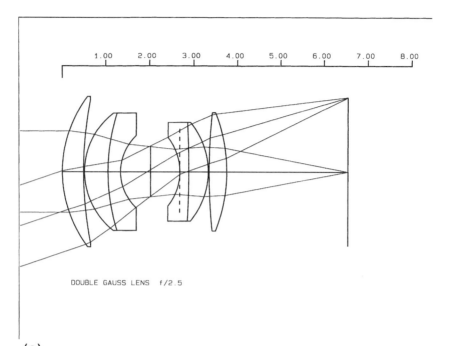

(a)

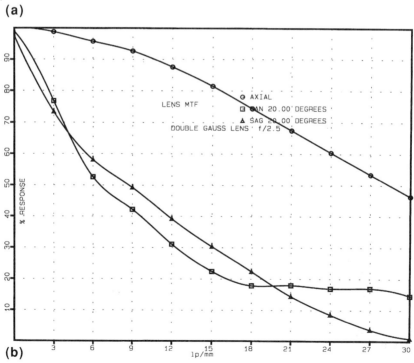

(b)

Figure 6-1 (a) Double Gauss lens $f/2.5$. (b) Lens MTF.

Double Gauss Lenses

Table 6-1 f/2.5 Double Gauss

	Radius	Thickness	Material	Diameter
0	0.00000	1.00000E+07	Air	
1	3.19354	0.4902	LAK9	3.680
2	9.50214	0.0226	Air	3.500
3	1.89471	0.5421	LAK9	2.880
4	5.17823	0.2827	SF8	2.880
5	1.28019	0.6849	Air	1.780
6	Stop	0.6663	Air	1.234
7	−1.43984	0.2009	F2	1.680
8	−25.74193	0.4418	LAK9	2.420
9	−2.01304	0.0202	Air	2.420
10	14.99510	0.4021	LAK9	2.900
11	−4.21828	2.7891	Air	2.900

The distance from the first lens surface to the image is 6.543. Distortion, at full field, is 0.3%.

Since this lens has vignetting, it is important that the designer adjust lens diameters so that the system is physically realizable. This is because most computer programs consider vignetting by appropriately adjusting the entrance pupil diameter at the various field positions to satisfy the designer's vignetting instruction. The lens diameters must be such that only those rays traced actually pass through the lens. It is never a problem if there is no vignetting; the aperture stop prevents this. Referring to the lens diagram, note how the first lens limits the lower rim ray at full field, while the last lens limits the upper rim ray.

Figure 6-3 shows an f/1 lens of focal length 4. The FOV is 5 degrees. It is a minor modification to what this author submitted at the 1980 Lens Design Conference (Juergens, 1980). In Table 6-3 is the lens data for Fig. 6-3.

Unfortunately, this lens has a few thick elements, a short back focal length, and no room for an iris mechanism. However, considering its low f#, it has excellent image quality. See also the projection lens example of Fig. 22-1.

In Fig. 6-4 is shown a 25 mm f/0.85 lens. This is a modification of that of Kitahara (1999) reviewed by Caldwell (1999). The front

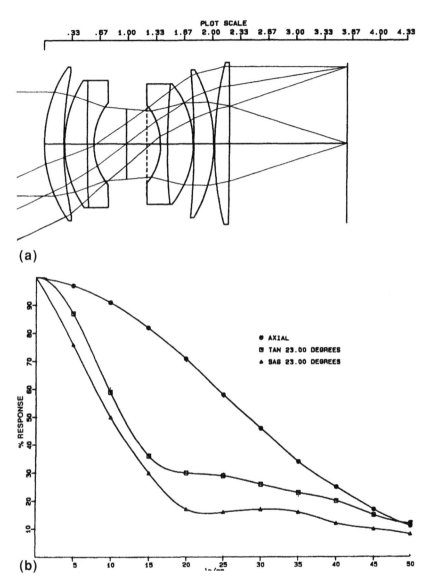

Figure 6-2 (a) 50 mm f/1.8 SLR camera lens. (b) Lens MTF.

Table 6-2 50 mm f/1.8 Lens for SLR

Radius	Thickness	Diameter	Material
	−1.212	1.12	Entrance pupil
1.3308	0.229	1.64	BASF2
3.3747	0.011	1.54	Air
1.1317	0.268	1.36	LAKN13
14.2879	0.080	1.36	SF2
0.6586	0.385	0.89	Air
IRIS	0.402	0.75	Air
−0.6248	0.081	0.84	SF8
5.6198	0.311	1.29	LAK8
−0.9558	0.011	1.29	Air
−8.5637	0.236	1.58	LAKN12
−1.4712	0.011	1.58	Air
3.0666	0.181	1.73	LAK8
−46.3791	1.398	1.73	Air

The distance from the first lens surface to the image is 3.604. This lens is a minor modification to that of Wakimoto (1971). The patent description covers an f/1.4 lens.

Table 6-3 4.0 EFL f/1 Lens

Radius	Thickness	Diameter	Material
	−5.132	4.00	entrance pupil
3.5519	1.399	4.43	LAF2
Flat	0.009	4.43	Air
1.7043	0.939	2.94	PSK3
−16.2381	0.252	2.94	SF1
0.9860	0.434	1.62	AIR
Stop	0.278	1.60	AIR
−1.6424	0.270	1.62	SF1
1.8028	1.159	2.05	LAF2
−2.3336	0.447	2.05	Air
1.6593	1.044	1.86	LAF2
14.4112	0.813	1.86	Air

The distance from the first lens surface to the image is 7.042. Distortion = 0.2%.

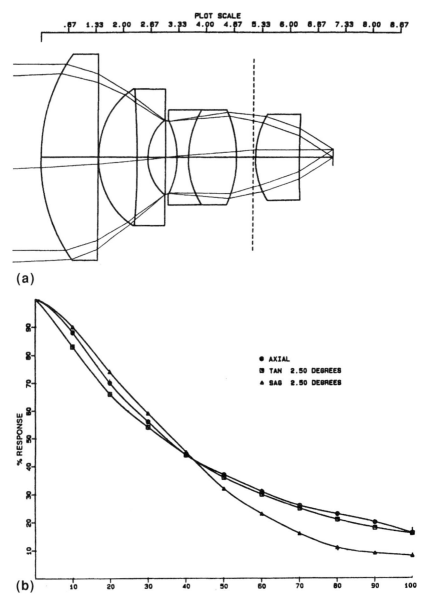

Figure 6-3 (a) $f/1$, 5° field of view, double Gauss lens. (b) Lens MTF.

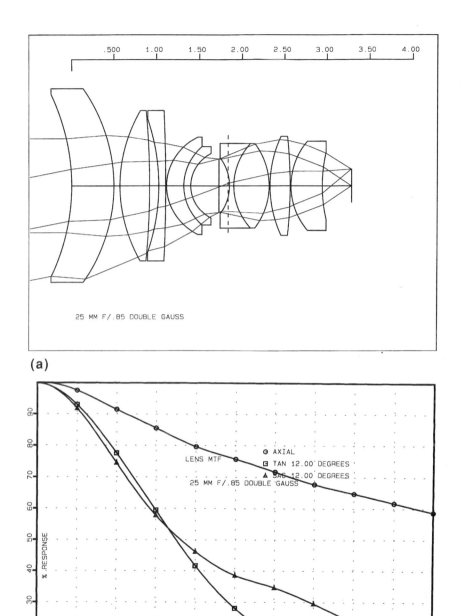

Figure 6-4 (a) 25 mm *f*/0.85 double Gauss lens. (b) Lens MTF.

Table 6-4 25 mm F/.85 Double Gauss Lens

	Radius	Thickness	Material	Diameter
	0.00000	1.00000E+07	Air	
1	−2.83155	0.5001	LASF36A	2.280
2	−2.19063	0.0700	Air	2.420
3	1.97886	0.3494	LASF3	1.880
4	−10.65614	0.1067	Air	1.880
5	−2.75248	0.0805	SF6	1.660
6	−21.99313	0.0100	Air	1.880
7	0.69612	0.2000	LASF3	1.220
8	0.67653	0.0806	SF6	0.980
9	0.46644	0.3260	Air	0.810
10	Stop	0.1256	Air	0.679
11	−0.60866	0.0507	SF57	0.700
12	0.81321	0.4156	LASFN31	1.060
13	−0.91272	0.0105	Air	1.060
14	1.55910	0.2367	LASFN31	1.240
15	−6.18794	0.0104	Air	1.240
16	0.90347	0.3699	LASF3	1.120
17	1.99074	0.3431	Air	0.840

meniscus element is acting as a coma corrector to the system. Unfortunately, there is an abundance of high-refractive-index glasses in this design. However, this is required for correction at large numeric apertures. In Table 6-4, is the lens data for this. Distortion is 1.2%. The distance from the first lens surface to the image is 3.286.

REFERENCES

Brandt, H. M. (1956). *The Photographic Lens,* Focal Press, New York.
Caldwell, B. (1999). Fast double Gauss lens, *Optics and Photonic News, 10:*38.
Fujioka, Y. (1984). Camera lens system with long back focal distance, U.S. Patent # 4443070.
Imai, T. (1983). Standard photographic lens system, U.S. Patent #4396255.
Juergens, R. C. (1980). The sample problem: a comparative study of lens de-

sign programs and users, 1980 International Lens Design Conference, *SPIE, 237*:348.
Kidger, M. J. (1967). Design of double Gauss systems, *Applied Optics, 6:*553.
Kitahara, Y. (1999). Fast double Gauss lens, U.S. Patent #5920436.
Mandler, W. (1980). Design of basic double Gauss lenses, *SPIE, 237:*222.
Mori, I. (1983). Gauss type photographic lens, U.S. Patent #4390252.
Mori, I. (1984). Gauss type photographic lens, U.S. Patent #4426137.
Momiyama, K. (1982). Large aperture ratio photographic lens, U.S. Patent #4364643.
Wakamiya, K. (1984). Great aperture ratio lens, U.S. Patent #4448497.
Wakimoto, Z. and Yoshiyuki, S. (1971). Photographic lens having large aperture, U.S. Patent #3560079.

7
Telephoto Lenses

Telephoto lenses consist of a front positive (F_a) and a rear negative (F_b) group separated by a distance T (see Fig.7-1) The telephoto ratio is defined as L/F (Cooke 1965). Assuming thin lenses we have

$$\frac{1}{F} = \frac{1}{F_a} + \frac{1}{F_b} - \frac{T}{F_a F_b}$$

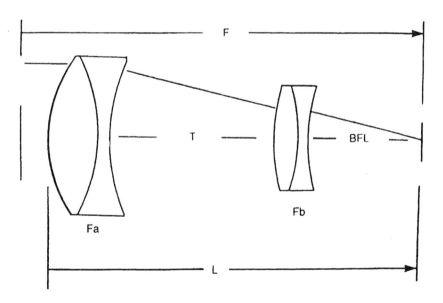

Figure 7-1 Telephoto lens.

$$BFL = \frac{F(F_a - T)}{F_a}$$

Solving this we obtain

$$F_a = \frac{TF}{F - L + T} \text{ and } F_b = \frac{T(T-L)}{F-L}$$

The absolute value of F_b is maximum at $T=L/2$.

In Fig. 7-2 is a telephoto lens derived from Aklin, 1945, f/5.6, with a 20 degree FOV and an EFL of 5. See Table 7-1.

The distance from the first lens surface to the image is 4.171. Distortion is 0.4% and the telephoto ratio is 0.83. The glass NLAF33 is a new glass in which the lead oxide has been replaced with titanium oxide. For manufacturing convience, the radii on surfaces 5 and 6 should be tied together to yield the same radii.

In Fig. 7-3 is shown an f/2.8 telephoto lens for use with a SLR camera (film size 1.703 diagonal). In Table 7-2 the lens prescription is given.

Distortion = 0.08%. The distance from the first lens surface to the image is 6.802. The telephoto ratio = 0.96.

Table 7-1 EFL = 5.0 Telephoto Lens

	Radius	Thickness	Material	Diameter
	0.00000	1.00000E+07	Air	
1	0.80002	0.2866	BK10	1.020
2	−2.18598	0.0758	BASF10	1.020
3	2.91954	0.0774	Air	0.840
4	Stop	0.9228	Air	0.776
5	−0.55672	0.0759	ZKN7	0.780
6	−0.55398	0.1005	NLAF33	0.840
7	−1.04609	0.0156	Air	0.980
8	34.63436	0.1365	SF1	1.070
9	−2.84343	2.4796	Air	1.070

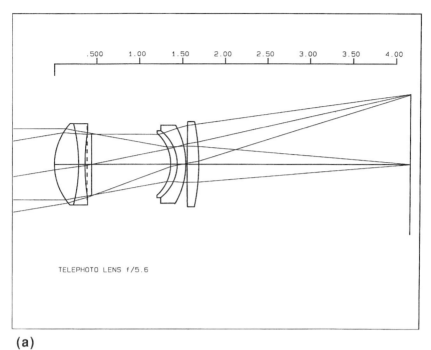

(a)

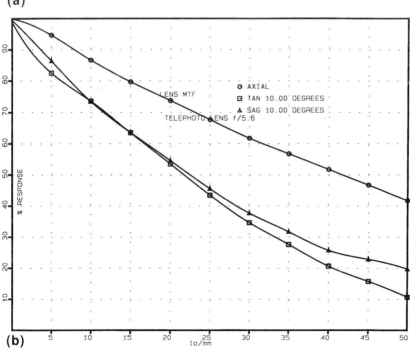

(b)

Figure 7-2 (a) Telephoto lens $f/5.6$. (b) Lens MTF.

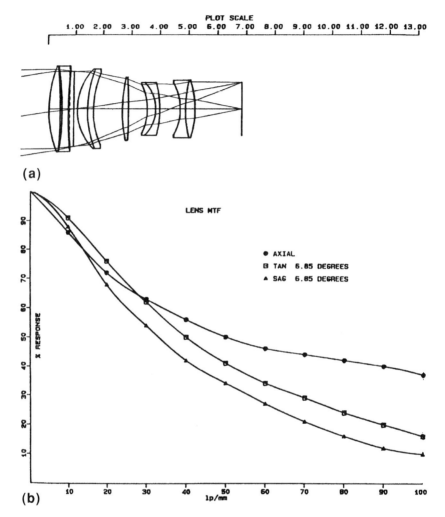

Figure 7-3 (a) $f/2.8$ 180 mm telephoto lens. (b) Lens MTF.

Telephoto Lenses 89

Table 7-2 f/2.8 180 mm Telephoto Lens

	Radius	Thickness	Material	Diameter
0	0.00000	1.00000E+07	Air	
1	3.30455	0.4174	PK2	2.730
2	−13.00895	0.0754	Air	2.730
3	−7.01395	0.2347	F5	2.730
4	11.92290	0.1827	Air	2.500
5	Stop	0.1204	Air	2.380
6	1.66033	0.3696	FPL53	2.540
7	2.99976	0.2152	SF1	2.540
8	2.56344	1.0264	Air	2.320
9	4.08741	0.2193	LLF6	2.020
10	−16.91399	0.6573	Air	2.020
11	−1.65427	0.2964	K10	1.600
12	−1.11837	0.1444	SF1	1.700
13	−2.90997	0.7049	Air	1.700
14	−1.62087	0.1774	ZK1	1.700
15	3.49063	0.3314	SF5	1.840
16	−2.76130	1.6287	Air	1.840

Comparing this design with the previous design we note that although this lens is not nearly as compact (it has larger telephoto ratio) as the design in Fig. 7-2, it has superior performance. This is made possible by the use of the Ohara FPL53 glass in the design. Although this design is not apochromatic, the use of this glass has greatly improved performance.

Internal focusing is generally used on relatively long lenses of this type. It has the mechanical advantage of keeping the entire lens fixed to the camera, while moving only an internal element. For examples of this see Kreitzer (1982) or Sato (1994). In this case, the next-to-last cemented doublet (K10 and SF1 glasses) is moved toward the film plane. Table 7-3 shows this focusing movement.

In Fig. 7-4 is shown a 400 mm focal length f/4 telephoto lens. It covers an image diameter of 1.069 inch and so is useful for 35 mm motion picture cinematography.

Table 7-3 Focusing Movement for 180 mm Telephoto Lens

Object distance	T(10)	T(13)
Infinity	0.657	0.705
384.52	0.717	0.645
194.56	0.777	0.585
131.25	0.837	0.525
99.62	0.897	0.465
80.65	0.957	0.405
68.01	1.017	0.345

Table 7-4 400 mm Telephoto Lens f/4

	Radius	Thickness	Material	Diameter
0	0.00000	1.00000E+08	Air	
1	10.93502	0.3166	FK5	4.200
2	−84.15616	0.0236	Air	4.200
3	3.37372	0.8165	FK51	3.980
4	−19.47681	0.6471	Air	3.980
5	−10.65171	0.3150	SF1	3.160
6	7.89857	0.3695	Air	2.900
7	Stop	2.3993	Air	2.766
8	3.83829	0.3912	SF1	2.020
9	−2.76429	0.2047	LAF3	2.020
10	1.69264	1.3484	Air	1.700
11	1.95739	0.2058	SF1	1.860
12	1.49895	0.2767	FK5	1.860
13	4.18140	4.8085	Air	1.700

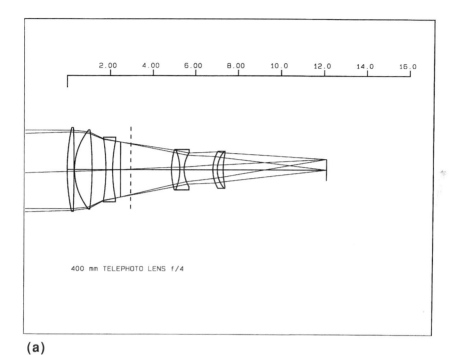

(a)

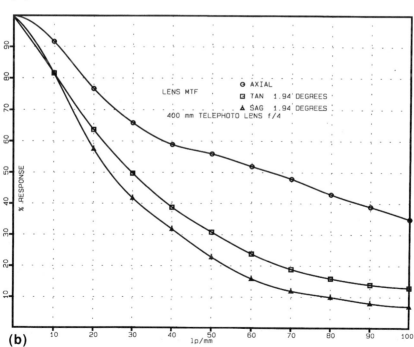

(b)

Figure 7-4 (a) 400 mm telephoto lens $f/4$. (b) Lens MTF.

Internal focusing may also be done (as in the above case) by moving the lens group after the stop. The distance from the first lens surface to the image is 12.123. Telephoto ratio = 0.77.

In Fig. 7-5 is shown a 1000 mm f/11 telephoto lens. It covers a diameter of 1.704 inch and so may be used for a single lens reflex (SLR) camera. The overall length is 19.720, and so its telephoto ratio is 0.5. The reader might find it instructive to compare this design to the Cassegrain lens of Fig. 15-4. The distortion is + 0.3% (pincushion).

Quite frequently, it is desirable to extend the focal length of an existing objective (F_a) by adding a negative achromat (F_b) a distance T from this objective. In addition to creating a telephoto lens, it may also be used as a dual mode system. That is, by inserting a mirror between the objective and the negative achromat, we may obtain a dual focal length system. Consider the cemented achromat in Fig. 2-2. If we want to double the focal length of this lens, then from the above equations with $T = 8$, a negative achromat of focal length of –24 is required.

Table 7-5 Apochromat Telephoto for SLR

	Radius	Thickness	Material	Diameter
0	0.00000	1.00000E+08	Air	
1	2.63238	0.3731	LAKN12	4.360
2	2.06779	1.7493	FPL53	3.800
3	–4.49150	0.3731	SK16	3.800
4	–9.97634	1.1447	Air	4.360
5	–15.42440	0.1976	SF1	2.240
6	48.62893	0.3277	Air	2.140
7	–2.29235	0.1980	LAK10	2.040
8	–13.07957	0.0948	Air	2.140
9	3.51306	0.5882	FPL53	2.100
10	2.93062	0.1713	Air	1.880
11	5.27029	0.1937	SF1	2.100
12	52.33926	1.4482	Air	1.880
13	Stop	2.6049	Air	1.383
14	–2.84691	0.1463	LAK10	1.220
15	3.23717	0.2169	F5	1.300
16	–3.38932	9.8926	Air	1.300

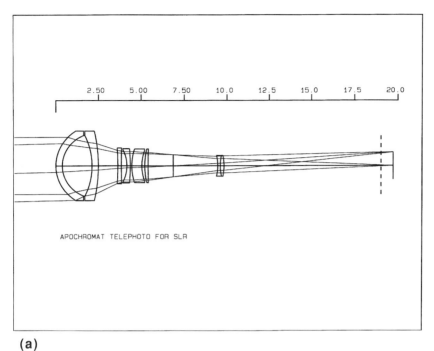

(a)

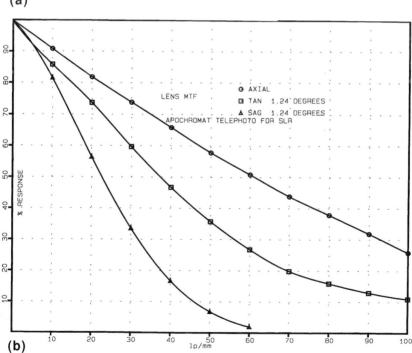

(b)

Figure 7-5 (a) Apochromat telephoto lens for SLR. (b) Lens MTF.

Using the same materials, and BAK1 and SF8 as the objective, we use the thin-lens achromat equations to arrive at a starting solution. (Refer to Table 2-1.) The system is then optimized and we vary only the parameters of the negative lens. Figure 7-6 shows the results of this. The effective focal length is 40 and the FOV = 1.5 degrees. In Table 7-6 is the lens data for Fig. 7-6.

For some glass combinations, as the front objective, a reversed form of the rear achromat is to be preferred.

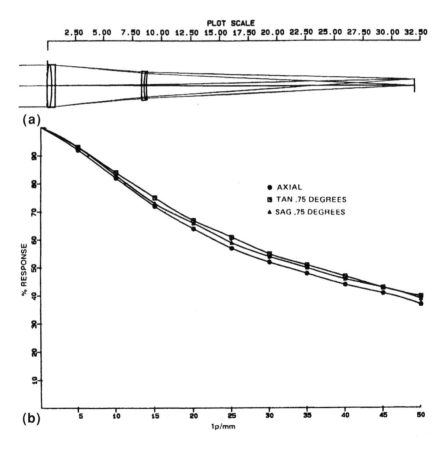

Figure 7-6 (a) Cemented achromat with 2x extender. (b) Lens MTF.

Table 7-6 Cemented Achromat with Extender

	Radius	Thickness	Material	Diameter
0	0.00000	1.00000E+08	Air	
1	12.80180	0.4343	BAK1	3.410
2	−9.06229	0.3208	SF8	3.410
3	−37.65534	7.5539	Air	3.410
4	15.62625	0.2301	BAK1	2.320
5	4.51190	0.1986	SF8	2.320
6	6.43392	23.3191	Air	2.320

The distance from the first lens surface to the image is 32.057. Distortion is nearly zero (due to the very small FOV).

REFERENCES

Aklin, G. (1945). Telephoto objective, U.S. Patent #2380207.
Arai, Y. (1984). Large aperture telephoto lens, U.S. Patent #4447137.
Cooke, G. H. (1965). *Photographic Objectives, Applied Optics and Optical Engineering,* R. Kingslake, ed., Academic Press, New York, Vol. 3, p. 104.
Eggert, J. (1968). Photographic telephoto lenses of high telephoto power, U.S. Patent #3388956.
Horikawa, Y. (1984). Telephoto lens system, U.S. Patent #4435049.
Kreitzer, M. H. (1982). Internal focusing telephoto len U.S. Patent #4359272.
Matsui, S. (1982). Telephoto lens system, U.S. Patent #4338001.
Sato, S. (1994). Internal focusing telephoto lens, U.S. Patent #5323270.
Tanaka, T. (1986). Lens system, U.S. Patent #4575198.

8
Inverted Telephoto Lenses

The inverted telephoto, or retrofocus, lens consists of a negative front lens group and a rear positive lens group. It is characterized by having a long BFL in relation to its effective focal length. This type of construction is extensively used in short-focal-length wide-angle lenses for SLR cameras. Here, a long BFL is required to clear the moving mirror and shutter mechanisms.

It is thus important when setting up the merit function in the optimization program properly to limit the minimum allowable back focal length.

Figure 8-1 shows an inverted telephoto. The requirements for this lens were that it have a BFL of at least 1.8 F (Laikin, 1974) Lens data for Fig. 8-1 is in Table 8-1.

Table 8-1 Inverted Telephoto Lens

	Radius	Thickness	Material	Diameter
0	0.00000	1.00000E+07	Air	
1	16.19458	0.0964	PK50	0.820
2	0.53435	0.5895	Air	0.650
3	0.57741	0.1178	SF1	0.820
4	0.53133	0.2701	Air	0.700
5	2.35781	0.3038	LF7	0.820
6	−1.51618	0.2158	Air	0.820
7	Stop	0.0740	Air	0.640
8	0.00000	0.0709	SF1	0.820
9	0.72364	0.2104	LAK21	0.760
10	−1.28454	2.1689	Air	0.760

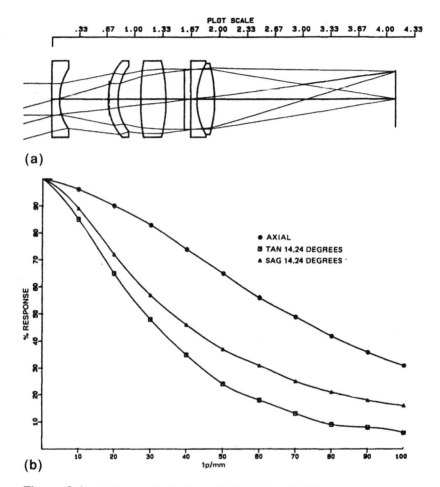

Figure 8-1 (a) Inverted telephoto *f*/3.5. (b) Lens MTF.

Inverted Telephoto Lenses

The distance from the front lens to the image is 4.118. The lens focal length is 1.181, f/3.5, visual region. Note that the lenses all have the same diameter and so may readily be assembled into a housing with the same inside diameter. With a little further work, the front surface probably could have been set plane.

The field of view is 28.5 degrees. Distortion is 2.6%. Note that the second element is a meniscus and nearly concentric about the aperture stop. It helps reduce spherical aberration of the assembly.

Figure 8-2 shows a wide-angle inverted telephoto lens. It is a modification of the Albrecht patent. It has a focal length of 10 mm, is f/2.8 with a 70 degree FOV.

The distance from the first lens surface to the image is 3.036. This lens shows relatively little chromatic aberration. The main aberration is oblique spherical. This causes the drop in tangential MTF response. At full field, distortion is 10.5% and there is 16% vignetting. However, since the entrance pupil is enlarged at the edge of the field, the relative illumination is 0.957 of the axial value.

Table 8-2 Wide-Angle Inverted Telephoto

	Radius	Thickness	Material	Diameter
0	0.00000	1.00000E+07	Air	
1	−3.40689	0.0874	LAF2	1.420
2	0.80610	0.4349	PSK3	1.210
3	−1.56736	0.0467	Air	1.210
4	1.52317	0.2421	SF3	0.980
5	−1.52317	0.0649	K7	0.980
6	0.29342	0.6220	Air	0.530
7	Stop	0.3013	Air	0.252
8	0.00000	0.0604	SF1	0.640
9	1.08515	0.1488	SK16	0.640
10	−0.86725	0.0153	Air	0.640
11	1.55722	0.1296	LAKN12	0.690
12	−1.42496	0.0154	Air	0.690
13	0.83187	0.1878	PSK3	0.670
14	−0.75273	0.0598	SF1	0.670
15	1.37683	0.6194	Air	0.590

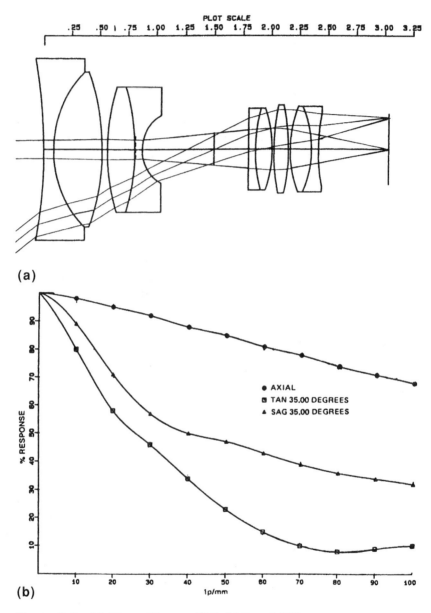

Figure 8-2 (a) 10 mm Cinegon f/2.8. (b) Lens MTF.

In Fig. 8-3 is shown an inverted telephoto lens suitable as a camera lens for 16 mm cinematography. This covers a diagonal of 0.5 inch and has a focal length of 0.628 (16 mm). The overall length is 3.355. Distortion is 2.1%. The field of view is 43.28 degrees. There is ample room for a shutter mechanism in front of the stop, so this lens may be used on a still camera. This lens is a modification of the Miles (1972) patent. See Table 8-3.

An interesting variation of the inverted telephoto design is its use as a soft focus photographic lens for portraiture. Since the portrait photographer does not want a lens with extreme resolving power that would show the subject's facial blemishes, a lens with spherical aberration is sometimes used. (Another trick is to use some grease on the front element to soften the image.) Caldwell (1999) discusses such a system in his review of the Hirakawa (1999) patent.

Table 8-3 Inverted Telephoto for a Camera $f/2.0$

Surf	Radius	Thickness	Material	Diameter
0	0.0000	0.100000E+11	Air	
1	1.3936	0.286	LAF2	1.48
2	−6.6020	0.118	LLF6	1.48
3	0.5106	0.933	Air	0.88
4	0.5237	0.092	SF5	0.60
5	0.3319	0.490	Air	0.50
6	STP	0.013	Air	0.52
7	2.0899	0.265	LAK10	0.68
8	−0.6855	0.046	Air	0.68
9	1.8935	0.197	K5	0.68
10	−0.4378	0.060	SF1	0.68
11	−1.1966	0.010	Air	0.68
12	0.8468	0.113	K5	0.58
13	−0.8097	0.079	F5	0.58
14	0.5542	0.654	Air	0.50

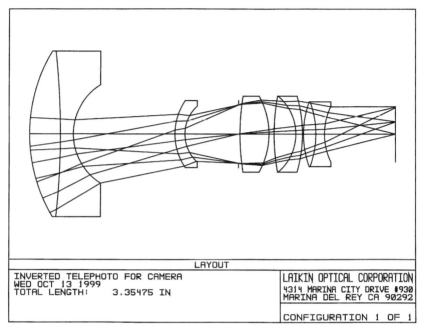

(a)

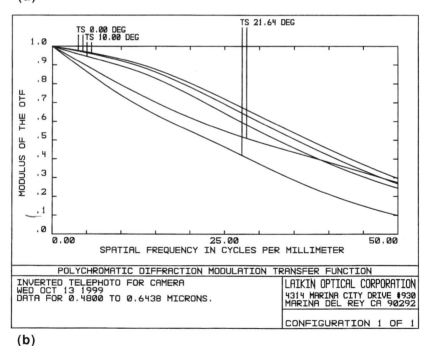

(b)

Figure 8-3 (a) Inverted telephoto camera lens. (b) Lens MTF.

REFERENCES

Albrecht, W. (1962). High speed photographic objective with wide image angle, U.S. Patent #3045547.

Caldwell, B. (1999). Wide angle soft focus photographic lens, *Optics and Photonics News, 10:* 49.

Cooke, G. H. (1955). Optical objectives of the inverted telephoto type, U.S. Patent #2724993.

Fujibayashi, K. (1980). Compact retrofocus wide angle objective, U.S. Patent #4235519.

Hirakawa, J. (1999). Wide angle soft focus photographic lens, U.S. Patent #5822132.

Hopkins, R. E. (1952). Wide angle photographic objective, U.S. Patent #2594021.

Hudson, L. M. (1960). Photographic objective, U.S. Patent #3064533.

Kingslake, R. (1966). The reversed telephoto objective, *SMPTE, 76:*203.

Kubota, T. (1979). Small retro-focus wide angle photographic lens, U.S. Patent #4134945.

Laikin, M. (1974). High resolution reverse telephoto lens, U.S. Patent #3799655.

Miles, J. R. (1972). Objective lens for short focal length camera, U.S. Patent #3672747.

Momiyama, K. (1984). Inverted telephoto type wide angle objective, U.S. Patent #4437735.

Mori, I. (1972). Retrofocus type lens system, U.S. Patent #3635546.

Tsunashima, T. (1979). Wide angle photographic objective, U.S. Patent #4163603.

9
Very Wide Angle Lenses

These lenses have fields of view of greater than 100 degrees and are of the inverted telephoto design. Such large field angles result in very large distortion, so large that we really should not even quote distortion as such. The lens of course has focal length, but this again has little meaning considering the large distortion.

If Y is the image height (measured from the optical axis to image centroid) and θ the semifield angle, then for a distant object $Y = F \tan(\theta)$ if there is no distortion.

In fisheye lenses, $Y =$ approximately $0.015\ F\theta$ (Laikin, 1980). That is, image height is nearly a linear function of the field angle.

With increasing field angle, the entrance pupil moves from inside the lens toward the front of the lens. This pupil movement is extremely apparent when examining such lenses. Pupil aberration must be considered during the design; ray starting data must be constantly readjusted at the various field angles. During analysis and MTF calculations, this starting data must be also adjusted at the various wavelengths.

During the preliminary design phase, determining the initial pupil shift values is a difficult task. To this end, the author has modified a ray trace program to trace a chief ray at some small field angle. The transverse height at the aperture is used to determine (by simple proportion) a new value at the entrance pupil. With a few iterations the actual chief ray is found. This pupil shift data is then fitted to a cubic equation to predict the starting value for the longitudinal shift of the next field angle. Increments of about two degrees are generally used.

All this is done automatically in slightly longer than it takes to print out the data.

In the ZEMAX program, this determination of pupil shift is done automatically.

Also the computer program should be such that the EFL is allowed to vary at will, distortion is ignored, and only the image heights at various field angles are maintained.

Due to the very short focal length, axial secondary color will not be a problem. However, the large field angle makes lateral secondary color impossible to eliminate.

Large distortion creates a unique problem; lens aberrations are sensitive to object distance changes. For convenience, this author generally designs projection lenses first at an infinite conjugate. Then, for the last few computer runs, finite conjugate and screen curvature is introduced. However, all the following examples are presented for an infinite conjugate and are corrected over the visual region.

Figure 9-1 shows a 100 degree FOV camera lens. It was designed to cover the 35 mm academy format (1.069 diagonal). The focal length is 0.73 and it is f/2.

Note from the ray diagram that this lens has a nearly telecentric exit pupil. As a camera lens, this is immaterial; however if this is to be used for projection, one must either place a negative field lens between the film gate and the arc source to locate properly the arc image (see the comments about exit pupil location under Example 9-3) or redesign this for proper exit pupil location.

This is an interesting design since only two common glasses are used and it does not contain any of the lanthanum types so frequently used in some of the other designs in this section.

Table 9-1 gives the paraxial location of the entrance pupil. From this value, the pupil is shifted, along the axis toward the front of the lens as in Table 9-2.

Figure 9-2 shows an f/2 120 degree FOV projection lens for 70 mm cinematography. The lens focal length is 1.22.

Table 9-3 gives the paraxial location of the entrance pupil for the lens shown in Fig. 9-2. From this value, the pupil is shifted, along the axis toward the front of the lens as in Table 9-4.

Very Wide Angle Lenses

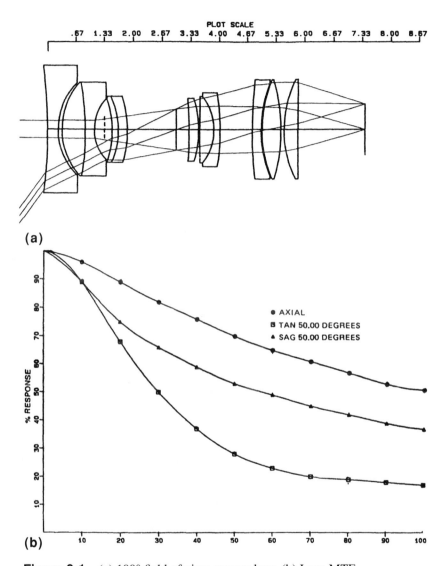

Figure 9-1 (a) 100° field of view camera lens. (b) Lens MTF.

Table 9-1 100 deg FOV Camera Lens

	Radius	Thickness	Material	Diameter
0	0.00000	1.00000E+07	Air	
1	−9.23618	0.2493	SK4	2.680
2	1.28960	0.0924	Air	1.960
3	1.38202	0.5414	SF1	1.980
4	−5.43835	0.2136	SK4	1.980
5	0.89262	0.4126	Air	1.300
6	−1.84275	0.1394	SF1	1.300
7	−1.52432	0.2066	SK4	1.400
8	−2.18352	1.1359	Air	1.400
9	Iris	0.2989	Air	0.875
10	−6.62446	0.2255	SK4	1.330
11	−1.97414	0.0213	Air	1.330
12	25.66874	0.3400	SK4	1.400
13	−1.03832	0.1463	SF1	1.400
14	−4.20190	0.7575	Air	1.570
15	8.25177	0.2069	SF1	2.100
16	2.41933	0.0210	Air	1.980
17	2.45181	0.4422	SK4	2.100
18	−3.42632	0.0712	Air	2.100
19	2.16988	0.3167	SK4	2.100
20	33.66520	1.5604	Air	2.100

The distance from the first lens surface to the image is 7.399.

Table 9-2 Pupil Shift and Compression

Angle	Shift	Compression	Image height
20 deg	0.069	0.928	0.248
35 deg	0.220	0.779	0.409
50 deg	0.470	0.556	0.532

Very Wide Angle Lenses

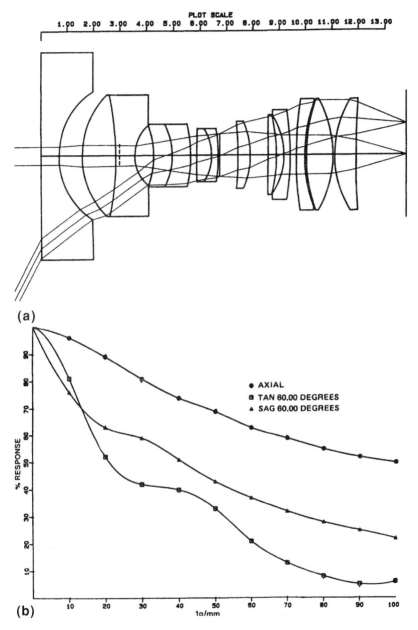

Figure 9-2 (a) 120° field of view projection lens. (b) Lens MTF.

Table 9-3 120 deg Projection Lens

	Radius	Thickness	Material	Diameter
0	0.00000	1.00000E+07	Air	
1	0.00000	0.6912	BAK4	6.950
2	2.43429	0.8803	Air	4.300
3	2.70448	1.2685	SF1	4.080
4	−8.65512	0.7206	SK4	4.080
5	1.25812	0.7045	Air	2.020
6	−2.76763	0.6868	SF1	2.020
7	−2.17607	0.6906	SK4	2.120
8	−4.38586	0.2195	Air	2.120
9	14.51959	0.5684	SK4	1.800
10	−1.55522	0.2486	SF1	1.800
11	−5.21620	0.0621	Air	1.800
12	Stop	0.6906	Air	1.450
13	−10.38549	0.4657	SK4	2.270
14	−2.58754	0.6637	Air	2.270
15	95.06452	0.3271	SK4	2.700
16	−5.07961	0.2841	Air	2.700
17	−2.20197	0.2424	SF1	2.650
18	−8.45074	0.2287	Air	3.030
19	11.44286	0.3148	SF1	3.750
20	4.63629	0.0653	Air	3.560
21	5.07851	0.9991	SK4	3.750
22	−3.49877	0.0612	Air	3.750
23	3.22557	0.8110	SK4	3.750
24	23.74864	1.9076	Air	3.750

The distance from the first lens surface to the image is 13.802.

Table 9-4 Pupil Shift and Compression for 120 deg Projection Lens

Angle	Shift	Compression	Image height
38 deg	0.419	0.813	0.748
45.8 deg	0.637	0.728	0.872
60 deg	1.182	0.534	1.052

Very Wide Angle Lenses

In Table 9-4 compression is the value of the tangential dimension of the entrance pupil as a fraction of the axial. The largest aberration with this lens is lateral secondary color (which is of course typical of all very-wide-angle lenses).

Figure 9-3 shows an f/2 160 degree FOV projection lens for 70 mm cinematography. Lens focal length = 0.900. This is a minor modification of a projection lens designed for Omni Flms Int. (Sarasota,FL) for their Cinema 180 theaters.

Table 9-5 160 deg FOV Projection Lens

	Radius	Thickness	Material	Diameter
0	0.00000	1.00000E+07	Air	
1	21.45628	0.5342	BK7	11.270
2	3.56551	2.0174	Air	6.420
3	0.00000	0.6296	BK7	6.620
4	5.38256	1.9296	Air	5.100
5	−5.38256	0.6296	BK7	5.100
6	0.00000	3.6842	Air	6.620
7	5.08416	0.5276	SF	2.400
8	−16.49561	0.0280	Air	2.400
9	2.30452	0.7150	SF4	2.400
10	2.22341	0.1560	Air	1.340
11	Stop	0.2876	Air	1.260
12	−1.96261	0.4509	SF8	1.380
13	2.54563	0.7439	LAKN7	2.200
14	−2.54563	0.0309	Air	2.200
15	11.68551	0.6871	LAKN7	2.380
16	−1.53496	0.2347	SF-4	2.380
17	−5.07431	0.0289	Air	2.660
18	3.38448	0.5704	LAKN7	2.780
19	30.24403	2.2006	Air	2.780

The distance from the front lens surface to the image is 16.086.

Table 9-6 Pupil Shift and Compression

Angle	Shift	Compression	Image height
40 deg	0.452	0.888	0.603
56 deg	0.965	0.771	0.809
80 deg	2.310	0.475	1.044

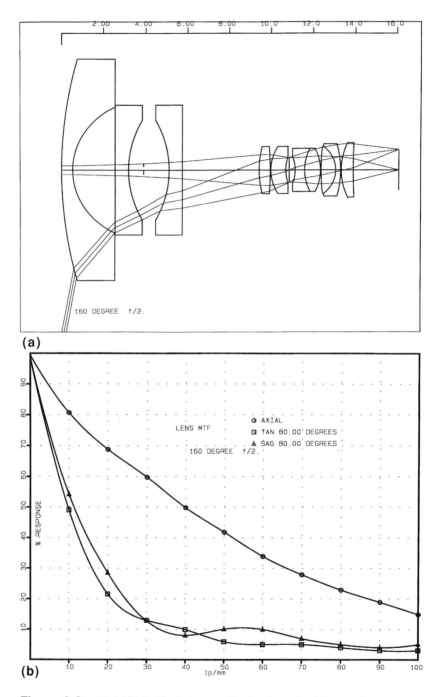

Figure 9-3 (a) 160° field of view projection lens for 70 mm cinematography (b) Lens MTF.

Very Wide Angle Lenses

Notice that the second and third lenses are really the same. Since this lens was designed for high-power Xenon arc projection (as were the lenses shown in Figs. 9-2 and 9-5), the following must be maintained (see also Chap. 22).

1. A BFL of at least 2 inches is required for film transport mechanism clearance.
2. Special optical cements must be used to withstand the high power densities.
3. The exit pupil should be about 4 to 6 inches in front of the film gate. The arc is imaged in this location for most projectors.

The reduced response at 80 degrees off axis as compared to axial is not really detrimental in wide-field projection systems. The observer's interest is mainly on the center of the screen and his visual acuity is greatly reduced at his peripheral vision.

The limits of the normal visual field are 160 degrees horizontal and 120 degrees vertical. Fields of vision greater than this require the observer to rotate his head (or entire body).

Figure 9-4 shows a 170 degree f/1.8 lens designed as a camera lens for 16 mm film (0.492 diagonal).

Note that the second surface is nearly a hemisphere. The computer program in use should have a means for preventing this surface from becoming a hyperhemisphere. In some programs the bound is the sag at the surface minus the semi-clear aperture at this surface. In ZE-MAX, one forms the product of the lens diameter and the curvature and then sets a limit to this operand.

Figure 9-5 shows a 210 degree FOV f/2 projection lens. It was designed to be used with a 10 perforation frame, 70 mm film. Therefore the 210 degree field corresponds to an image circle of 1.85.

The distance from the first lens surface to the image is 15.891. The entrance pupil is shifted from its paraxial location toward the front of the lens as in Table 9-10.

Axial, 60 degrees off-axis, and 105 degrees off-axis rays are traced. Note the extreme compression of the tangential component of the entering 105 degree rays.

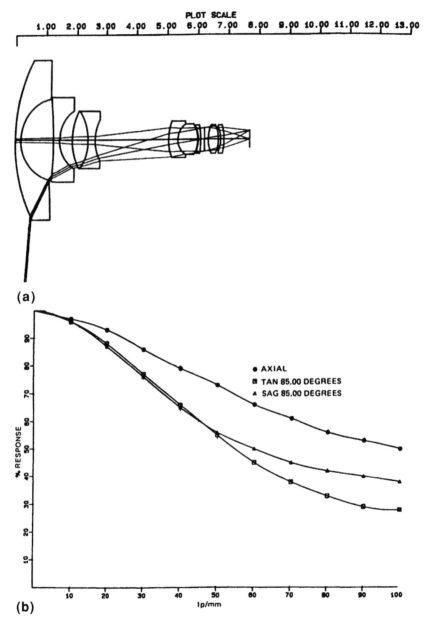

Figure 9-4 (a) 170° $f/1.8$ camera lens. (b) Lens MTF.

Very Wide Angle Lenses

Table 9-7 170 degree, $f/1.8$ Camera Lens

	Radius	Thickness	Material	Diameter
0	0.00000	1.00000E+07	Air	
1	4.97794	0.1796	BK7	4.650
2	1.19204	1.0797	Air	2.350
3	−10.73562	0.1939	SK5	2.250
4	0.97296	0.4213	Air	1.680
5	1.70682	0.5228	SF1	1.660
6	−1.38378	0.2399	LAK9	1.660
7	1.13444	2.4304	Air	1.200
8	1.37528	0.2991	LAK9	1.060
9	0.51233	0.6489	PSK3	0.880
10	−0.79167	0.0205	Air	0.880
11	−0.77634	0.0799	SF1	0.760
12	−1.90832	0.1200	Air	0.820
13	Iris	0.1207	Air	0.670
14	0.96375	0.3331	PSK3	0.820
15	−0.89273	0.0792	Air	0.820
16	−0.73648	0.1076	SF1	0.690
17	−1.42818	0.8308	Air	0.820

The distance from the first lens surface to the image is 7.707. The lens focal length is 0.2.

Table 9-8 Pupil Shift and Compression

Angle	Shift	Compression	Image height
34	0.088	0.900	0.115
59.5	0.299	0.739	0.189
85	0.760	0.497	0.246

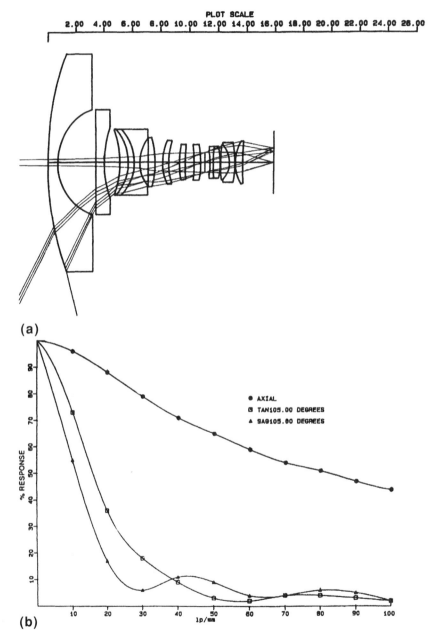

Figure 9-5 (a) 210° field of view projection lens. (b) Lens MTF.

Table 9-9 210 deg FOV Projection Lens

	Radius	Thickness	Material	Diameter
0	0.00000	1.00000E+07	Air	
1	18.53034	0.6999	K5	13.700
2	3.42384	2.7086	Air	6.570
3	0.00000	0.5999	K5	6.570
4	6.04912	1.1030	Air	4.500
5	−5.39443	0.5892	K5	4.050
6	−2.90775	0.4097	SF1	4.150
7	−2.82644	0.3769	PK2	4.150
8	2.20673	0.6103	Air	3.000
9	0.00000	0.4885	SF3	3.100
10	−5.11554	0.5160	Air	3.100
11	2.95828	0.4185	SSKN5	2.760
12	3.74636	0.7503	Air	2.600
13	5.14042	0.3987	SF3	2.260
14	8.98951	0.6366	Air	2.260
15	−7.49724	0.4585	SF5	2.260
16	−5.30447	0.2918	Air	2.260
17	Stop	0.3532	Air	1.270
18	−4.06978	0.2042	SF8	1.520
19	6.90124	0.5040	LAKN7	2.020
20	−3.81584	0.0200	Air	2.020
21	8.67629	0.6256	LAKN7	2.200
22	−1.51779	0.3819	SF-3	2.200
23	−5.93765	0.0239	Air	2.520
24	2.45072	0.5055	LAKN7	2.720
25	16.23018	2.2162	Air	2.720

Table 9-10 Pupil Shift and Compression

Angle	Shift	Compression	Image height
30 deg	0.234	0.921	0.385
60	1.049	0.692	0.700
85	2.527	0.387	0.867
105	4.538	0.101	0.924

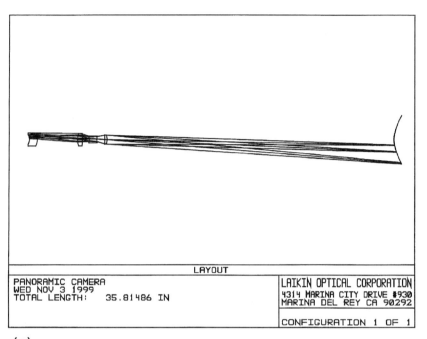

(a)

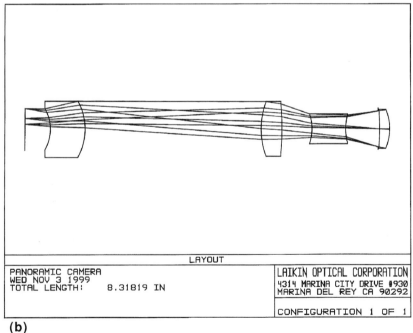

(b)

Figure 9-6 (a) Panoramic camera system. (b) Panoramic lens. (c) Len MTF.

Very Wide Angle Lenses 119

Figure 9-6a shows a panoramic camera system. The first surface shown (surface 2 in the prescription table) is a hyperbolic mirror with a conic coefficient of –2.7880 The ecentricity $\varepsilon = \sqrt{2.7880} = 1.6697$. Figure 9-6b shows just the lens portion of this.

Surface 1 is a reference surface. The lens of SF1 (surface 6) is a little too thick and could be made thinner. This forms a 360 degree panoramic image in the form of a donut. The outer diameter is 0.944 inch, corresponding to objects 60 degrees off axis. The inner diameter of this donut image is 0.29 inch, corresponding to objects 20 degrees off axis. The efective focal length of this system is 0.3099 inch and is f/6.85.

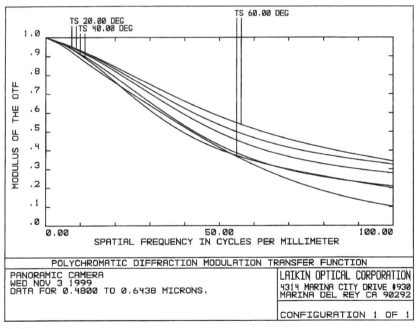

(c)

Figure 9-6

Surf	Radius	Thickness	Material	Diameter
0	0.0000	1.000E+11	Air	
1	0.0000	−7.4044	Air	18.987
2	3.0398	−28.0849	MIRROR	4.649
3	−1.1073	−0.2300	LAKN7	0.875
4	−6.1073	−0.0320	Air	0.811
5	Stop	−0.7390	Air	0.800
6	1.2583	−0.7643	SF1	0.615
7	−1.0051	−0.6869	Air	0.666
8	−5.5022	−0.4769	LAKN7	1.115
9	1.6652	−4.0005	Air	1.219
10	−1.2396	−0.8003	LAK21	1.249
11	−0.8902	−0.5882	Air	0.917

REFERENCES

Brewer, S., Harris, T., and Sandback, I. Wide angle lens system, U.S. Patent #3029699.

Chahl, I. S. and Srinivasan (1997). Reflective surfaces for panoramic imaging, *Applied Optics, 36:*8275.

Horimoto, M. (1981). Fish eye lens system, U.S. Patent #4256373.

Hugues, E. (1969). Wide angle short photographic objective, U.S. Patent #3468600.

Laikin, M. (1980). Wide angle lens system, *SPIE, 237:*530 (1980 Lens Design Conference).

Miyamoto, K. (1964). Fish-eye lens, *JOSA, 54:*1060.

Momiyama, M. (1983). Retrofocus type large aperture wide angle objective, U.S. Patent #4381888.

Muller, R. (1987). Fish eye lens system, U.S. Patent #4647161.

Shimizu, Y. (1973). Wide angle fisheye lens, U.S. Patent #3,737,214.

10
Eyepieces

An eyepiece is a lens system with an external entrance pupil. At this entrance pupil location the eye should be placed. The distance from the entrance pupil to the first lens surface is called eye relief. Eye relief should be at least 10 mm (to obtain adequate clearance for eyelashes); 15 mm to provide more comfortable viewing; 20 mm for those people wearing spectacles. In the case of eyepieces used for rifle sights, this eye relief should be at least 3 inches to allow for rifle recoil; see Figs. 14-4 and 35-20.

The iris of the eye varies from 2 mm in diameter (in bright sunlight) to 8 mm in diameter (dark viewing). Its focal length is approximately 17 mm (Luizov, 1984). This iris diameter D (in mm) is approximated by the empirical expression

$$D = 5.3 - 0.55 \ln B$$

See also Eq.(3) of Alpern, 1987, where B is the scene brightness in ft lamberts.

Some representative values of B are as in Table 10-1.

Table 10-1 Scene Brightness vs. Eye Pupil Diameter

Environment	Scene brightness (ft lambert)	D (mm)
Clear night	.01	7.8
Dawn/dusk	1	5.3
Office	100	2.7
Sunny day	1000	1.5

Typical eyepieces, then, are designed to cover the 3 to 6 mm diameter entrance pupil. If the viewer is observing in a moving vehicle (as in military equipment), entrance pupils as large as 10 mm in diameter are sometimes used. This is to prevent the observer from losing his field of view as his head moves from side to side.

The human eye can resolve (Ogle, 1951) 1 minute of arc (central foveal cone vision), about 3 minutes of arc at 5 degrees off axis, and 10 minutes of arc 20 degrees off axis (rod vision). This is for relatively high illumination levels, about 100 ft lamberts. Visual acuity drops as the illumination level decreases.

When designing a binocular device, be sure to allow sufficient interpupillary adjustment. This is typically 51 to 77 mm (or could be as low as 45 mm for some small children).

The difference in binocular magnification (aniseikona) should be kept to within 0.5%. The two systems should be aligned (disparity) as follows:

horizontal	8 minutes of arc convergent
	4 minutes of arc divergent
vertical	4 minutes of arc

Eyepieces should have the ability to move longitudinally (focus) thru 4 diopters. Newton's formula is

$$F^2 = -X X'$$

where X is the apparent image distance, and X' is the amount of eyepiece movement in inches (see Fig. 10-1), and D is the equivalent movement in diopters.

In most eyepieces, the distance from the eye to the principal plane is approximately a focal length. Therefore, the eyepiece forms an image, a distance X, as shown in Fig. 10-1, from the eye.

$$F^2 = \frac{39.37}{D} X'$$

The lens moves away from the eye for negative diopters, forming a virtual image in front of the eye.

Eyepieces

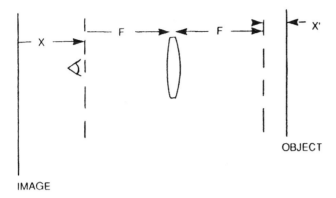

Figure 10-1 Positive diopter with focusing eyepiece.

The eyepiece magnification M is the ratio of the apparent image size as viewed through the eyepiece to the object size. Since this object is generally imaged at a distance of about 10 inches from the eye (the distance of most distinct vision, and thus the distance that one normally holds his reading material),

$$M = \frac{F_e + 10}{F_e}$$

where F_e is the eyepiece effective focal length.

When an eyepiece is combined with a microscope objective (to form a compound microscope), the resultant magnification is the product of objective magnification and the eyepiece magnification. However, when used with an objective F_o to form an afocal telescope, the angular magnification of this system is

$$M = \frac{F_o}{F_e}$$

Since the general design procedure is to trace from long conjugate to short, what is actually the system exit pupil (where the eye would be) is now called the entrance pupil.

The designer must then bound the exit pupil to correspond to the telescope objective, relay, microscope objective, etc. location. This is typically a positive distance of 5 to 30 inches. Some wide-angle systems can have considerable pupil aberration and it is then necessary to find the paraxial location of this exit pupil as well as using a chief ray to locate this pupil.

Figure 10-2 shows a 10× eyepiece with a 5 mm diameter entrance pupil and a 40 degree FOV. Distortion is 3.2%. It is a modification of the usual Kellner form. The exit pupil is located 32.45 from the last surface.

Figure 10-3 shows a 10× eyepiece with the same pupil diameter and field of view as above (see Konig, 1940). However, an element has been added and eye relief increased. Distortion is 2.8%. The exit pupil is located 32.54 from the last surface.

Table 10-2 10× Eyepiece

	Radius	Thickness	Material	Diameter
0	Pupil	0.9160	Air	0.200
1	4.54506	0.3677	SK5	1.060
2	−0.85817	0.0238	Air	1.060
3	1.10780	0.3488	PK2	0.980
4	−0.68646	0.1192	SF1	0.980
5	8.87539	0.5853	Air	0.900

The distance from the first lens surface to the image is 1.445.

Table 10-3 10×Eyepiece

	Radius	Thickness	Material	Diameter
0	Pupil	0.9870	Air	0.200
1	−5.36474	0.2682	PK2	1.080
2	−1.19131	0.0210	Air	1.080
3	3.16992	0.2684	PK2	1.140
4	−1.20795	0.0209	Air	1.140
5	1.02345	0.3585	PSK3	1.040
6	−0.80733	0.1359	SF1	1.040
7	1.25388	0.4709	Air	0.800

The distance from the first lens surface to the image is 1.544.

Eyepieces 125

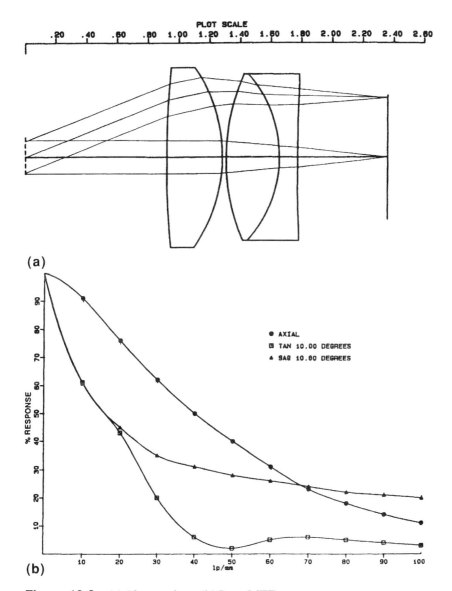

Figure 10-2 (a) 10× eyepiece. (b) Lens MTF.

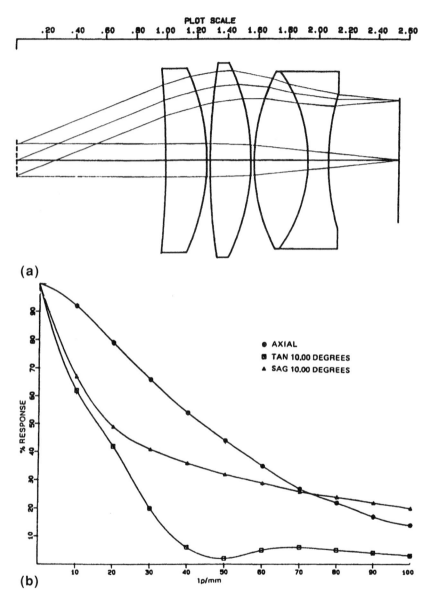

Figure 10-3 (a) 10× eyepiece. (b) Lens MTF.

Eyepieces

This type of design is slightly superior to the Plossl or symmetric type of eyepiece. This is shown in Fig. 10-4. It consists of two identical cemented doublets with the crown elements facing each other. It was designed for the same pupil diameter and field of view as the above two designs.

In the above three designs, ray plots show the rays at full field, but MTF is only for half field. This is due to the poor image quality at the full field.

Figure 10-5 shows an Erfle eyepiece with a 5 mm diameter en-

Table 10-4 Plossl Eyepiece

	Radius	Thickness	Material	Diameter
0	Pupil	0.761	Air	0.200
1	2.12177	0.1182	SF1	0.980
2	0.90425	0.2783	SK14	0.980
3	−1.77007	0.0197	Air	0.980
4	1.77007	0.2783	SK14	0.980
5	−0.90425	0.1182	SF	0.980
6	−2.12177	0.7250	Air	0.980

The distance from the first lens surface to the image is 1.538. Distortion is 6.5%. The exit pupil is 32.49 from the last surface.

Table 10-5 Erfle Eyepiece

	Radius	Thickness	Material	Diameter
0	Pupil	0.7050	Air	0.197
1	−0.85871	0.0900	SF1	0.780
2	2.64172	0.5085	SK14	1.340
3	−0.93336	0.0207	Air	1.340
4	2.40312	0.4450	LAK10	1.730
5	−1.73284	0.0210	Air	1.730
6	1.06666	0.5837	FK3	1.400
7	−0.98995	0.1993	SF1	1.400
8	1.48913	0.4971	Air	1.140

128 Chapter 10

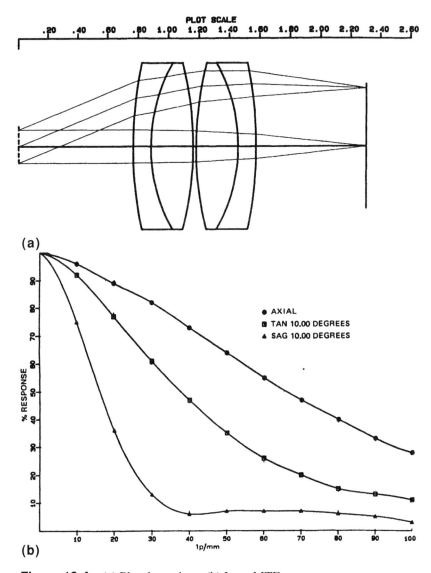

Figure 10-4 (a) Plossl eyepiece. (b) Lens MTF.

Eyepieces 129

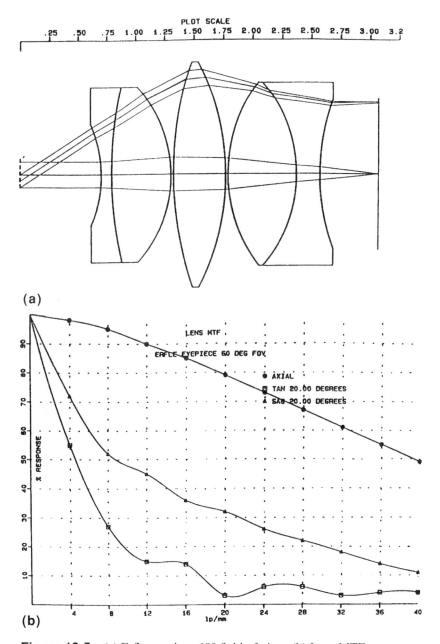

Figure 10-5 (a) Erfle eyepiece 60° field of view. (b) Lens MTF.

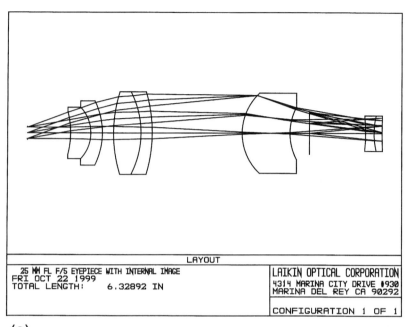

(a)

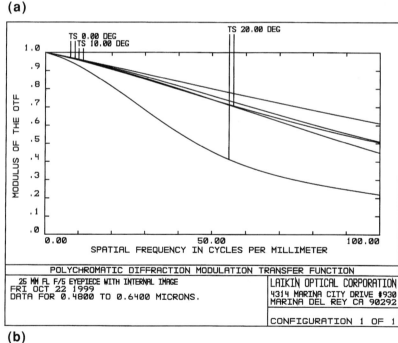

(b)

Figure 10-6 (a) 25 mm f/5 eyepiece with an internal image. (b) Lens MTF.

Eyepieces

Table 10-6 Eyepiece with Internal Image

Surf	Radius	Thickness	Material	Diameter
0	0.0000	0.100000E+11		0.00
1	STP	0.772		0.20
2	−1.4463	0.3686	LAKN22	0.740
3	−0.6331	0.2067	SF1	0.900
4	−1.2377	0.1980		1.120
5	2.3997	0.4924	ZKN7	1.440
6	−1.5105	0.2017	SF1	1.440
7	−2.4327	1.5941		1.440
8	0.9460	0.8486	LAKN22	1.400
9	0.9786	1.3533		0.920
10	−2.5607	0.1502	SK16	0.540
11	1.3128	0.1433	SF1	0.620
12	11.1380	−1.2983		0.620
13	0.0000	0.0000		0.730

trance pupil and a 60 degree FOV. Distortion is 4.9%. The exit pupil is located 21.93 from the last surface.

The distance from the first lens surface to the image is 2.365. In this design, ray plotting is for full field, but the MTF is plotted at 20 degrees off axis. This lens was optimized using four field angles: axial, 10°, 20°, and 30°. As can be seen from Fig. 10-5a, 50% vignetting was introduced and the pupil shifted so that for the full field case, only the bottom half of the pupil is traced.

Figure 10-6 shows an eyepiece with an internal image. This is a modification of the Nagler (1981) patent. It has an internal image, so its use is limited to systems where a reticle or grid is not required. It is also quite long (6.329 from the entrance pupil to the last lens surface). Note that this has the same focal length (1.0) and field of view (40 degrees) as the first three designs in this chapter. However, it has much superior performance. Distortion is less than 0.1%.

REFERENCES

Abe, H. (1971). Wide angle eyepiece. U.S. Patent #3586418.
Alpern, M. (1987). Eyes, and Vision, Section 12 of *Handbook of Optics*, Optical Society of America, McGraw-Hill, New York.

Andreyev, L. N. (1968). Symmetric eyepieces with improved correction, *Sov. J. Opt. Tech., 5:*303.

Bertele, L. (1929). Occular, U.S. Patent #1699682.

Clark, T. L. (1983). Simple flat field eyepiece, *Applied Optics, 22:*1807.

Fedorova, N. S. (1980). Relation between MTF and visual resolution of a telescope, *Sov. J. Opt.Tech. 47:*1.

Fukumoto, S. (1996). Eyepiece, U.S. Patent #5,546,237.

Giles, M. K. (1977). Aberration tolerances for visual optical systems, *JOSA, 67:*634.

MIL HDBK 141 (1962). Chapter 14, Optical Design, Military Standardization Handbook, Defense Supply Agency.

Koizumi, N. (1997). Wide field eyepiece with inside focus, U.S. Patent #5,612,823.

Konig, A. (1940). Telescope eyepiece, U.S. Patent #2206195.

Ludewig, M. (1953). Eyepiece for optical instruments, U.S. Patent #2637245.

Luizov, A. V. (1984). Model of reduced eye, *Sov. J. Opt. Tech., 51:*325.

Nagler, A. (1981). Ultra wide angle flat field eyepiece, U.S. Patent #4286844.

Nagler, A. (1987). Ultra wide angle eyepiece, U.S. Patent #4,747,675.

Olge, K. N. (1951). On the resolving power of the human eye, *JOSA, 41:*517.

Repinski, G. N. (1978). Wide angle five lens eyepiece, *Sov. J. Opt. Tech., 45:*287.

Rosen, S. (1965). Eyepieces and Magnifiers, Chapter 9, Volume 3, *Applied Optics and Optical Engineering,* Academic Press.

Skidmore, W. H. (1967). Eyepiece design providing a large eye relief, *JOSA, 57:*700.

Skidmore, W. H. (1968). Wide angle eyepiece, U.S. Patent #3390935.

Taylor, E. W. (1945). The evolution of the inverting eyepiece, *J. Scientific Inst., 22:*43.

Veno, Y. (1996). Eyepice, U.S. Patent #5,557,463.

Wald, G. (1945). The spectral sensitivity of the human eye, *JOSA, 35:*187.

Wenz, J. B. (1989). Single eyepiece binocular microscope, U.S. Patent #4,818,084.

11
Microscope Objectives

These are essentially diffraction limited optical devices. These lenses are characterized by specimen magnification and NA. The image is generally about 16 mm in diameter.

Most designs are based upon the old concept of a 160 mm tube length. That is, the distance from the image to a principal plane in the objective was 160 mm. This corresponds to an object-to-image distance of about 180 mm. They were designed to be mounted on a precision shoulder 35.68 mm from the specimen, which was covered by a 0.18 mm thick glass. This cover glass is made from either fused quartz (UV type objectives) or a chemically resistant soda lime glass. Since this type of glass is reasonably close to K-5, it is used in these designs. If there is oil immersion, then an additional thickness of 0.14 mm of oil is placed between the cover glass and the first lens surface (generally a plane surface).

The objective thread is a Whitworth and has a 55 deg included angle, 36 threads/inch, and a 0.796 major diameter. Typical laboratory microscopes have a shoulder-to-specimen distance of 45 mm.

Several manufacturers have instituted infinity-corrected systems. This allows greater flexibility in microscope design. Various beamsplitters and other accessories can now be readily inserted into the collimated beam. All designs shown here are for 180 mm object-to-image distance with a 16 mm diameter image.

According to the Raleigh criterion, the distance between two resolved images is $Z=0.61\lambda/NA$ where Z is the separation of the images, λ is the wavelength of light, and $NA = N \sin \theta$. N is the refractive index of the medium in which is the image. Thus to increase resolution we

Table 11-1 10× Microscope Objective

	Radius	Thickness	Material	Diameter
0	Stop	0.0119	Air	0.295
1	0.51518	0.1387	K5	0.390
2	−0.36923	0.0459	F2	0.390
3	−4.41903	0.3404	Air	0.390
4	0.45211	0.1323	K5	0.330
5	−0.23655	0.0383	F2	0.330
6	−0.83668	0.2967	Air	0.330
7	0.00000	0.0068	K5 cover glass	0.063
8	0.00000	0.0000		

The distance from the first lens surface to the image is 0.999. The object distance to the first lens surface is 6.076. The last surface represents the protective cover slip. Distortion = 0.26%.

need to decrease the wavelength of the illumination system (see design 11-4, a UV objective) or increase the *NA* beyond 1. This is done by immersing the object in oil (see design 11-5, a 98× oil imersion objective).

In Fig. 11-1 is shown a 10× microscope objective. It consists of two widely spaced cemented doublets (Lister type). The NA = 0.25 and the EFL = 0.591.

In Fig. 11-2 is shown a 20× microscope objective. The NA = 0.5 and the EFL = 0.328.

This lens has some primary color. Image quality is excellent on axis with the response falling off considerably at the edge.

Figure 11-3 shows a 4 mm FL apochromatic objective of NA = 0.95. Visual region. (Actually the focal length is 3.605 mm.)

Note that the next-to-last lens surface ($R = 0.09842$) is nearly concentric about the image. The spherical contribution at this surface is then zero. Distortion = 0.5%.

Magnification is 47.5. The 2.62 degree field angle shown in the MTF plot corresponds to a 16 mm diameter image. Resolution in the central part of the field is nearly diffraction limited. However, at the edge of the field, the design shows considerable lateral color. This is often solved by compensating for this in the eyepiece design.

Microscope Objectives

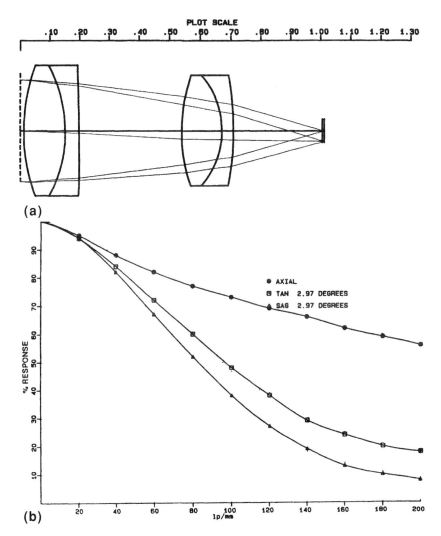

Figure 11-1 (a) 10× microscope objective. (b) Lens MTF.

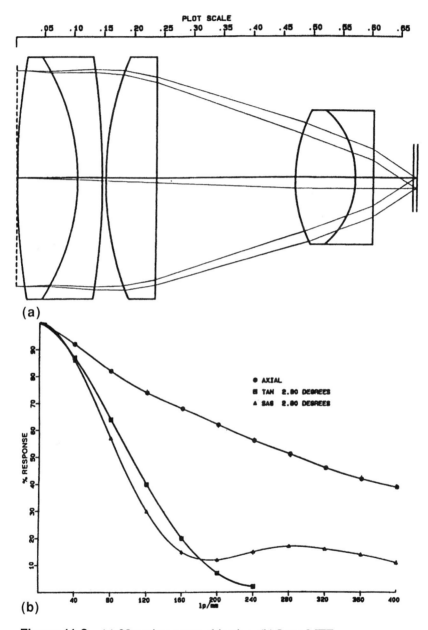

Figure 11-2 (a) 20× microscope objective. (b) Lens MTF.

Microscope Objectives

Table 11-2 20× Microscope Objective

	Radius	Thickness	Material	Diameter
0	Stop	0.0008	Air	0.320
1	0.90753	0.1027	K10	0.360
2	−0.30270	0.0413	SF1	0.360
3	−1.11024	0.0057	Air	0.360
4	0.42185	0.0854	SK5	0.360
5	Flat	0.2333	Air	0.360
6	0.19915	0.1017	SK	0.200
7	−0.12326	0.0309	F5	0.200
8	Flat	0.0681	Air	0.200
9	Flat	0.0070	K5	0.032

The distance from the object to the first lens surface is 6.410. The distance from the first lens surface to the image = 0.676. Distortion = 0.12%.

Table 11-3 4mm FL Apochromatic Objective

	Radius	Thickness	Material	Diameter
0	0.00000	6.5833	Air	0.630
1	0.37307	0.0258	LAKN12	0.320
2	0.17458	0.1361	CaF_2	0.260
3	−0.16802	0.0258	LAKN12	0.260
4	−0.62359	0.0099	Air	0.320
5	0.41731	0.0606	PSK3	0.320
6	−0.68979	0.0067	Air	0.320
7	Stop	0.0067	Air	0.260
8	0.16865	0.1072	CaF_2	0.240
9	−0.17242	0.0220	F1	0.240
10	−0.39327	0.0099	Air	0.240
11	0.09842	0.0773	SK5	0.180
12	0.00000	0.0083	Air	0.180
13	0.00000	0.0070	K5	0.013

The distance from the first lens surface to the image is 0.503.

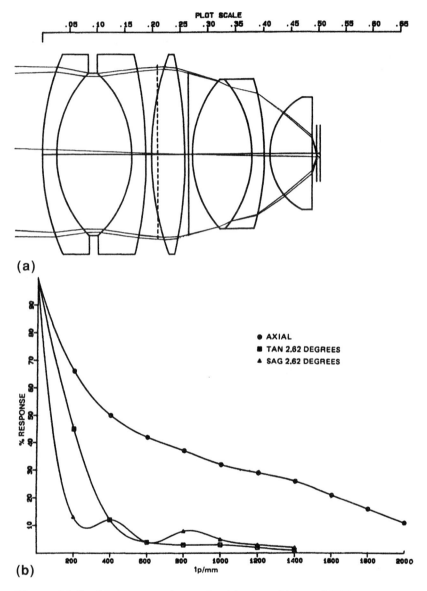

Figure 11-3 (a) 4 mm apochromatic objective. (b) Lens MTF.

Figure 11-4 shows a UV reflecting objective (UV spectral region, 0.2 to 0.4μm), NA = 0.72, magnification = 53, EFL = 0.130.

Table 11-4 UV Reflective Objective

	Radius	Thickness	Material	Diameter
0	Ent. pupil	−1.1545	Air	0.184
1	2.48604	0.0350	Quartz	0.360
2	7.53409	0.9731	Air	0.360
3	0.18197	−0.4995	Air-reflect	0.160
4	0.72977	0.5776	Air-reflect	1.100
5	2.87713	0.0427	CaF_2	0.600
6	0.48327	0.0016	Air	0.450
7	0.23165	0.1861	CaF_2	0.400
8	−1.27837	0.0011	Air	0.400
9	−1.20449	0.0588	Quartz	0.310
10	0.65590	0.0769	Air	0.180
11	0.00000	0.0071	Quartz	0.012

The distance from the object to the first lens surface is 5.6259. The distance from the first lens surface to the image is 1.4606. The aperture stop is located at the secondary mirror, which is bonded to the calcium fluoride negative lens.

Near diffraction-limited performance is obtained in the central portion of the field. At the edge of the field, coma limits the resolution. Since chromatic aberration is very small for this type of design, the wavelength region may be extended into the visual region.

The last surface in the above designs corresponds to the protective cover slip. Distortion = 0.2%. See Fig. 15-8 for a reflecting objective design that can be modified for microscope use.

Figure 11-5 shows a 98× oil immersion microscope objective of NA = 1.3 and EFL = 0.0711.

This type of design is sometimes called a semiapochromatic or fluorite objective. To be a true apochromat, a triplet should be added in the front. This cemented triplet would have as a middle element calcium fluoride. The oil is a microscope immersion oil (Cargille, 1995) and has the values for the index of refraction as in Table 11-1.

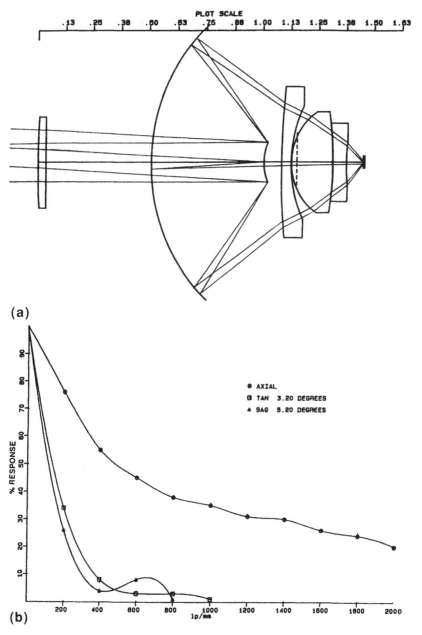

Figure 11-4 (a) UV-reflecting microscope objective 53×. Lens MTF.

Microscope Objectives

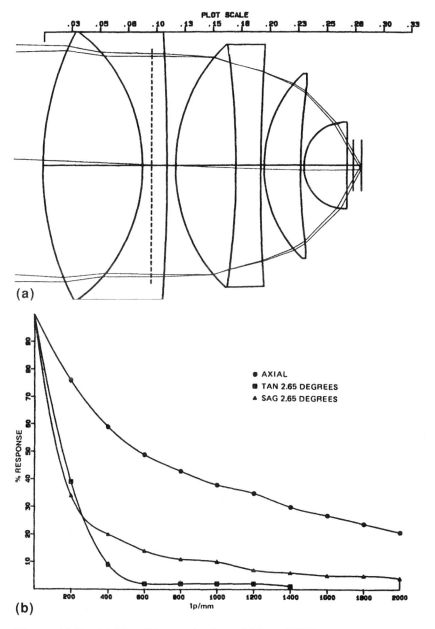

Figure 11-5 (a) 98× oil immersion lens. (b) Lens MTF.

Table 11-5 98× Oil Immersion Objective

	Radius	Thickness	Material	Diameter
0	Ent. pupil	−0.0957	Air	0.183
1	0.21035	0.0879	CaF_2	0.210
2	−0.12523	0.0216	F2	0.210
3	−1.50811	0.0073	Air	0.210
4	0.12314	0.0535	CaF_2	0.190
5	−0.60262	0.0215	F2	0.190
6	1.20979	0.0031	Air	0.190
7	0.09339	0.0321	SK14	0.145
8	0.40165	0.0030	Air	0.130
9	0.03356	0.0382	BK7	0.067
10	0.00000	0.0060	Oil	0.049
11	0.00000	0.0067	K5	0.029

The object's distance to the first lens surface is 6.8058. The distance from the first lens surface to the image is 0.2809.

Table 11-6 Refractive Properties of Type A Immersion Oil

Wavelength (μm)	Index of refraction
0.4861	1.5239
0.5461	1.5180
0.5893	1.5150
0.6563	1.5115

As can be seen, this closely approximates the refractive properties of BK-7 (at the e line). The last element then is made of BK-7 and is made as a hyperhemisphere. The front plano side is not antireflection coated, since it will be in contact with the immersion oil. Such hyperhemispheres are sometimes made by a ball rolling process as are ball bearings. That is, these small parts are not blocked as in conventional lens manufacture but are rolled between two parallel plates. A near perfect sphere is then obtained, and then these parts are conventionally blocked and the flat side ground and polished. Distortion = 0.5 %.

The main off-axis aberration is coma, which severely limits performance at the edge of the field. However, for the typical clinical application (examination of blood cells, etc.) the primary interest is in the central portion of the field.

REFERENCES

Beck, J. L. (1969). A new reflecting microscope objective, *Applied Optics,* 8:1503.
Benford, J. (1965). Microscope Objectives, Chapter 4 of *Applied Optics and Optical Engineering,* Vol. 3, Academic Press, New York.
Benford, J. (1967). Microscopes, Chapter 2 of *Applied Optics and Optical Engineering,* Vol. 4, Academic Press, New York.
Cargille, R. P. (1995). Microscope immersion oils data sheet, IO-1260, Cargille Labs, 55 Commerce Rd, Cedar Grove, N.J. 07009.
Esswein, K. (1982). Achromatic objective, U.S. Patent #4362365.
Grey, D. S. and Lee, P. H. (1949). A new series of microscope objectives, *JOSA, 39:*727.
Martin, L. C. (1966). *Theory of the Microscope,* Blackie and Son, London, 1966.
Matsubara, M. (1977). Microscope objective, U.S. Patent #4037934.
Rybicki, E. (1983). 40× microscope objective, U.S. Patent #4379623.
Sharma, K. (1985). Medium power micro objective, *Applied Optics, 24:*299.
Sharma, K. (1985). High power micro objective, *Applied Optics, 24:*2577.
Suzuki, T. (1997). Ultra high NA microscope objective, U.S. Patent #5,659,425.
Sussman, M. (1983). Microscope objective, U.S. Patent #4376570.

12
In-Water Lenses

Visibility in water is limited by scattering. Indeed, there are many places in the world where a bather cannot see his own feet! Twenty feet in seawater could be regarded as the limit of practical seeing. Figure 12-1 gives the transmission of 10 meters of seawater (from Smith and Baker, 1981). If we keep in mind that most optical glasses show considerable absorption at wavelengths below 0.4 microns, a reasonable wavelength region for optimization would be 0.4 to 0.55 microns (this should also be modified by the illumination and detector system sensitivity). However, designs discussed in this section are for the normal visual region.

Using the Schott equation (also called a Laurent series formula) for refractive index N,

$$N^2 = F_1 + 0.01F_2 \lambda^2 + 0.01F_3 \lambda^{-2} + F_4(10\lambda)^{-4} + F_5(10\lambda)^{-6} + F_6(10\lambda)^{-8}$$

where λ is the wavelength in microns (note that the factors 0.01 and 10 have been added to the usual form of this equation to avoid leading zeros in the coefficients).

Values for F_1, F_2, etc. for pure water (Centino, 1941) and seawater were fitted by a least squares fit (20 deg C).

This yields the following refractive index values:

Huibers (1997) also lists index of refraction values for seawater and pure water.

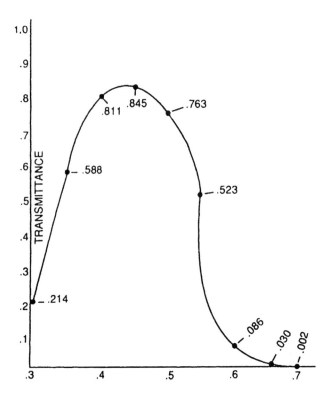

Figure 12-1 Transmittance through 10m of ocean water.

Table 12-1 Coefficients in Dispersion Formula

	Pure water	Sea water
F_1	1.76650	1.78364
F_2	−1.48402	−1.49409
F_3	0.44545	0.43052
F_4	3.90462	4.63499
F_5	−16.30518	−21.55772
F_6	41.34407	51.64058

In-Water Lenses

Table 12-2 Refractive Index Values for Water

Wave length (μ)	Pure water	Sea water
0.40	1.34308	1.34973
0.45	1.33914	1.34569
0.50	1.33638	1.34287
0.55	1.33434	1.34078
0.60	1.33275	1.33916
0.65	1.33145	1.33785

The increase in refractive index of seawater is due to salt concentration. It increases approximately 0.00185/1% salt concentration. The above values assume a salt concentration of 3.5%, which is the average salinity of the world ocean (3.5 grams of salt per 100 grams of water; oceanography texts refer to this concentration as 35‰).

The index of refraction of water decreases 0.0001/deg C at the above salt concentration and over the temperature region 10 to 20 deg C (Quan, 1995).

In general, for in-water systems, if F' is the distance from the second principal plane to the image of a distant object and F a like distance measured from the first principal plane, then for an image in air,

$$F' = \frac{F}{N}$$

where N is the refractive index of the water.

Image height will be (in the absence of distortion) $NF \tan\theta$. The actual optical system used depends upon the type of port required, which is usually a flat plate or a concentric dome.

FLAT PLATE PORT

The simplest type of port is, of course, a flat plate. This is what we encounter when we look at fish in an aquarium or use the usual face mask in a pool. Objects in water appear larger than they actually are. Figure 12-2 shows the relationship between an angle in water and its

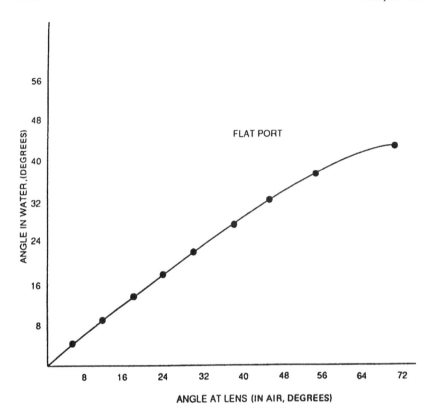

Figure 12-2 Angle in water versus angle in air for a flat port of 0.5-in. Pyrex.

exit angle in air after passing through a 0.5 inch thick Pyrex® flat port. For small angles the relationship is nearly linear and so there is no distortion, only a change in field of view. For large angles, the view in water is compressed, and so we note distortion. Since all of this is wavelength dependent, we will see considerable lateral color: it is very noticeable on viewing fish (at large angles of incidence) in an aquarium.

Figure 12-3 shows an optical system designed for in-water photography using a flat port camera and 70 mm film (this is a modification of a design by R. Altman).

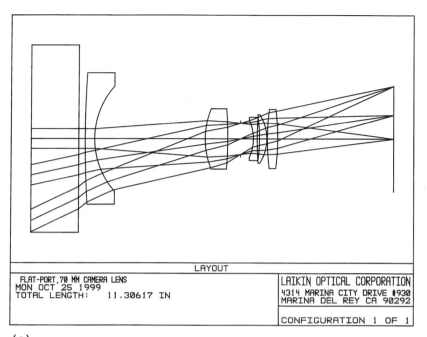

(a)

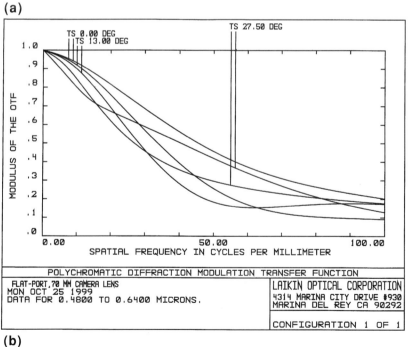

(b)

Figure 12-3 (a) Flat-port, 70 mm camera lens. (b) Lens MTF.

Table 12-3 Flat Port 70 mm Camera Lens

Surf	Radius	Thickness	Material	Diameter
0		1.000E+10	Seawater	
1	Flat	1.5000	Pyrex®	5.700
2	Flat	0.2000	Air	5.700
3	30.0125	0.3000	SK16	4.000
4	2.3720	3.4344	Air	3.200
5	2.3500	0.7040	LAKN7	1.840
6	−11.6351	0.4022	Air	1.840
7	Iris	0.397	Air	0.951
8	−1.3367	0.1205	SF8	1.140
9	9.3756	0.0702	Air	1.320
10	−4.7790	0.2331	LAKN12	1.320
11	−1.3442	0.0186	Air	1.480
12	7.0017	0.3226	LAKN13	1.800
13	−6.6768	3.6035	Air	1.800
14	0.0000	0.0000		3.220

PYREX® is a trade name of Corning Glass. The entrance pupil is located a distance of 4.800 from the first surface of the window.

This lens is f/3.9, has a focal length of 2.372, and was designed for a camera using 70 mm film and having an image diagonal of 3.3. Distortion is 2.6%. FOV is 55 deg in water. Distance from the front of the Pyrex® port to the image surface is 11.306.

MTF at mid-field is actually a little worse than full-field.

CONCENTRIC DOME

For very deep water exploration the concentric dome is preferred. This is based upon the enormous pressure requirements that one encounters in deep water (say 5000 feet, where the pressure is 2200 lbs/sq in.). Concentric dome ports of acrylic material are generally chosen over glass because of their reliability. Figure 12-4 shows a water–dome interface.

If point A is the center of curvature of the dome and is also the

In-Water Lenses

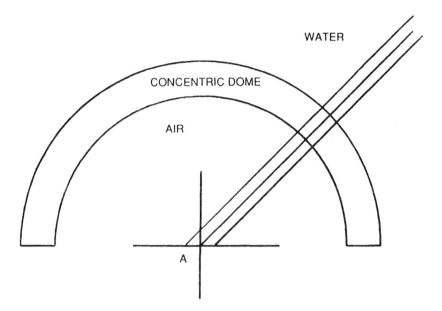

Figure 12-4 Water-dome interface.

entrance pupil of the system, then a chief ray from a distant object will pass through the water–dome interface without refraction. If the entrance pupil is kept small compared with the radius of curvature of the dome, then upper and lower rim rays will also pass through with only a slight change in their angles.

Since a concentric lens has its principal planes at its center of curvature, if the lens has its front principal plane also at this location, the net effect is a system with the same field of view in water and in air (Pepke, 1967).

To compensate for the aberrations of the water–dome interface and to move the pupil to a more accessible location, a corrector is sometimes used. Since the water–dome interface acts like a strong negative lens, considerable difficulty is also encountered in focusing a standard lens placed behind this interface. The afocal corrector solves these problems. This will allow a stock lens, designed for infinity in air correction, to be used behind this corrector. The system shown in Fig. 12-5

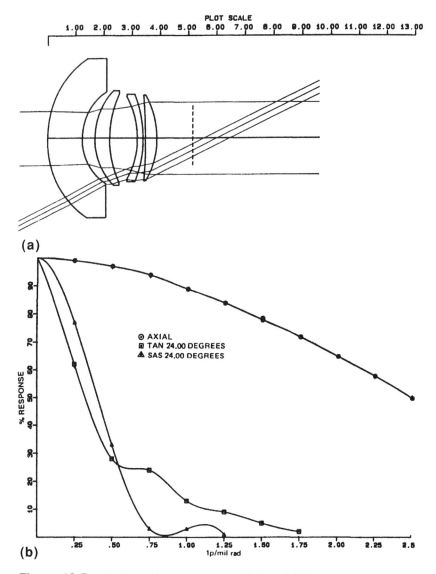

Figure 12-5 (a) Water-dome corrector. (b) Lens MTF.

Table 12-4 Water-Dome Corrector

	Radius	Thickness	Material	Diameter
0	0.00000	1.00000E+07	Seawater	
1	2.90001	1.2500	Acrylic	5.000
2	1.65000	0.4475	Air	2.900
3	2.00100	0.5218	FK5	2.700
4	3.01500	0.9755	Air	2.600
5	−2.45000	0.2000	BASF10	2.600
6	−3.40400	0.0850	Air	2.700
7	−14.65045	0.4017	BK10	2.600
8	−2.50700	5.6990	Air	2.700

has unit angular magnification and was designed to be used with a 6-to-1 zoom lens.

The field of view is 48 degrees. The entrance pupil is 5.145 from the first surface of the dome. The exit pupil is 2.039 inches from the last lens surface. The distance from the front of the acrylic dome to the exit pupil is 5.92. The focal length of the lenses behind the dome is 5.729. Distortion is very low, 0.2%.

Note: The entrance pupil diameter has been reduced from its axial value of 1.7 to 0.283 at 24 degrees off axis. This is to match the pupil diameter of the zoom lens as it goes from narrow field to wide field. A lens placed behind this afocal corrector sees its field of view in air the same as in water. However, the entrance pupil diameter has increased by a ratio equal to the index of refraction of the water.

In Fig. 12-6a an f/2.8 lens suitable for a 35 mm single lens reflex camera (43.3 mm diagonal) is shown. This is a modification of Ohshita (1989). Note that the dome is not strictly concentric. It has some negative power. This forms a retrofocus system, thus allowing a long back focal length in relation to its effective focal length (1.119). Focusing is accomplished by moving the rear four elements in relation to the dome.

The distance from the first lens surface to the image is 4.020. Distortion is 1.3%. The entrance pupil is 1.6715 from the first surface of the dome. MTF data is shown in Figure 12-6b.

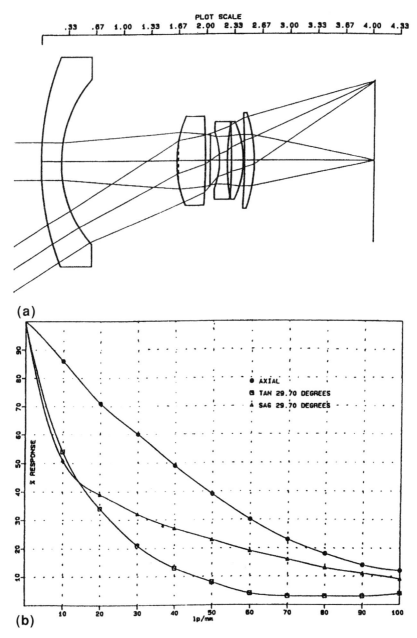

Figure 12-6 (a) In-water lens with dome. (b) Lens MTF.

Table 12-5 In-Water Lens with Dome

	Radius	Thickness	Material	Diameter
0	0.00000	1.00000E+07	Seawater	
1	2.72436	0.2435	BK7	2.230
2	1.22947	1.4244	Air	1.760
3	1.06980	0.3256	LAFN21	0.940
4	−5.97491	0.0558	Air	0.940
5	Stop	0.1091	Air	0.540
6	−0.78030	0.0919	SF4	0.620
7	1.92244	0.0534	Air	0.740
8	−6.13483	0.1388	LAFN21	0.740
9	−0.85199	0.0170	Air	0.860
10	−5.72057	0.1132	LAFN21	0.900
11	−1.41092	1.4470	Air	0.980

REFERENCES

Centeno, M. (1941). Refractive index of water, *JOSA, 31:*245.

Defant, A. (1961). *Physical Oceanography,* Chapter 2, Vol. 1, Pergamon Press.

Hale, G. M. and Querry, M. R. (1973). Optical constants of water, *Applied Optics, 12:*555.

Huibers, P. D. T. (1997). Models for the wavelength dependence of the index of refraction of water, *Applied Optics, 36:*3785.

Hulbert, E. O. (1945). Optics of distilled and natural waters, *JOSA, 35:*698.

Jerlov, N. G. (1968). *Optical Oceanography,* Elsevier, New York.

Jerlov, N. G. and Nielsen, R. S. (1974). *Optical Aspects of Oceanography,* Academic Press, New York.

Mertens, L. (1970). *In-Water Photography,* Wiley Interscience, New York.

Ohshita, K. (1989). Photo taking lens for an underwater camera, U.S. Patent #4,856,880.

Padgitt, H. R. (1965). Lens system having ultra-wide angle of view, U.S. Patent #3175037.

Palmer, K. and Williams, D. (1974). Refractive index of water in the IR, *JOSA, 64:*1107.

Pepke, M. H. (1967). Optical system for photographing objects in a liquid medium, U.S. Patent #3320018.

Quan, X. and Fry, E. S. (1995). Emperical equation for the index of refraction of seawater, *Applied Optics, 34:*3477.

Smith, R. and Baker, K. (1981). Optical properties of the clearest natural waters, *Applied Optics, 20:*177.

SPIE Seminar Proceedings (1968). Underwater photo-optical instrumentation, SPIE, Bellingham, WA.

SPIE Seminar Proceedings (1986). *Ocean Optics,* SPIE 637; Palmer, K. and Williams, D. (1974). Refractive index of water in the IR, *JOSA, 64:*1107.

13
Afocal Optical Systems

Afocal optical systems are used in power changers in microscopes, IR scanning or FLIR systems, laser beam expanders, viewfinders, and so on.

Some computer programs simply use a very large image distance for ray trace evaluation, while others use a "perfect lens" at the exit pupil. This converts angular errors to intercept errors so that one can obtain the usual plots of intercept and path length errors, spot diagrams and MTF analysis.

Figure 13-1 shows a 5× laser (HeNe, 0.6328 µm) beam expander. There is an intermediate image 0.396 to the right of the first plano surface. At this location is generally placed a small aperture for mode selection.

Table 13-1 5× Laser Beam Expander

	Radius	Thickness	Material	Diameter
0	0.00000	1.00000E+07	Air	
1	0.29627	0.0620	SF8	0.200
2	Flat	2.4812	Air	0.200
3	Flat	0.1320	SF8	0.700
4	−1.48084		Air	0.700

The overall length of the lens is 2.675.

Note that the MTF data presented here in Figs. 13-1, 13-2, and 13-3, as well as the other designs for laser systems, assumes a uniformly illuminated entrance pupil. Since these laser systems have a

157

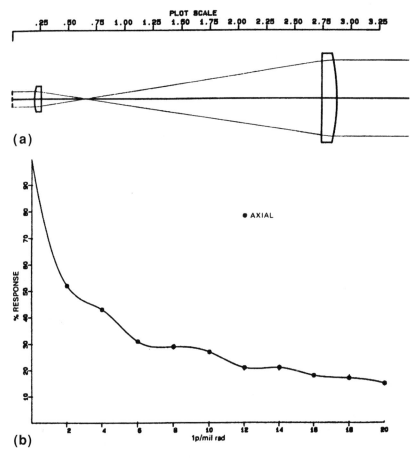

Figure 13-1 (a) 5× laser beam expander. (b) Lens MTF.

Gaussian distribution, the MTF response should be slightly higher than that shown here

Since there are here two essentially thin lenses, separated by a large air space, as a preliminary design, we shape each lens for minimum spherical aberration. For simplicity, a plano convex form was chosen. Since the refractive index for minimum spherical aberration is 1.686, SF8 glass is chosen (1.68445). With this type of all-positive lens, spherical aberration adds.

Afocal Optical Systems 159

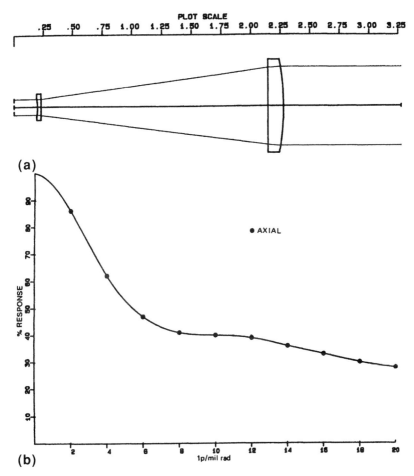

Figure 13-2 (a) 5× Galiean beam expander. (b) Lens MTF.

Using a negative–positive system (a Galilean telescope) greatly reduces spherical aberration and makes a more compact system. However, a mode selection aperture is not possible. On the other hand, for a very high power laser (not this HeNe laser), it might be preferred, since there is no focused spot and so less chance of the ionization of air.

The Galilean system is shown in Fig. 13-2.

Table 13-2 5× Galilean Beam Expander

	Radius	Thickness	Material	Diameter
0	0.00000	1.00000E+07	Air	
1	−0.34429	0.0300	SF8	0.150
2	Flat	1.9154	Air	0.200
3	Flat	0.1320	SF8	0.700
4	−1.72112		Air	0.700

The overall length of the lens is 2.077.

Both systems were designed for a 3 mm diameter entrance pupil.

Figure 13-3 shows a Galilean type beam expander of 50× magnification. It was designed to accommodate a 2 mm diameter input HeNe (0.6328 μm) laser beam.

Table 13-3 50× Laser Beam Expander

	Radius	Thickness	Material	Diameter
0	0.00000	1.00000E+07	Air	
1	−0.18905	0.0590	F2	0.090
2	0.69752	11.0680	Air	0.180
3	−15.13200	0.3360	SF8	3.900
4	−7.85965	0.0390	Air	4.100
5	Flat	0.3694	SF8	4.100
6	−16.11744		Air	4.100

The overall length of the lens is 11.871.

Due to the large amount of spherical aberration present, the positive group was split into two lenses. In order to provide more negative spherical aberration for the negative component, a glass of lower refractive index was used, F2 (1.61656). Probably, additional improvement could be made by changing this to an ordinary crown (BK7).

In figure 13-4 is shown a simple Galilean 4× plastic telescope. Since the exit pupil for such systems lies inside the telescope, a positive value of 0.5 (corresponding to a 12 mm eye relief) was nevertheless used. At 2 degrees off axis the pupil is then vignetted by 50% and shifted by half the entrance pupil. That is, we trace for only the lower half of the entrance pupil. Two of these units are often used to form a low cost binocular. The overall length of the lens is 3.709. Following is the data for Fig. 13-4.

Afocal Optical Systems 161

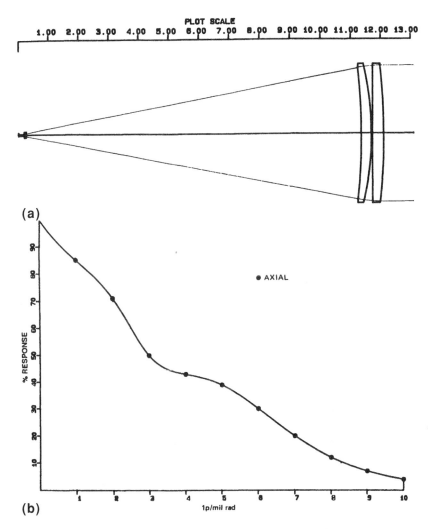

Figure 13-3 (a) 50× laser beam expander. (b) Lens MTF.

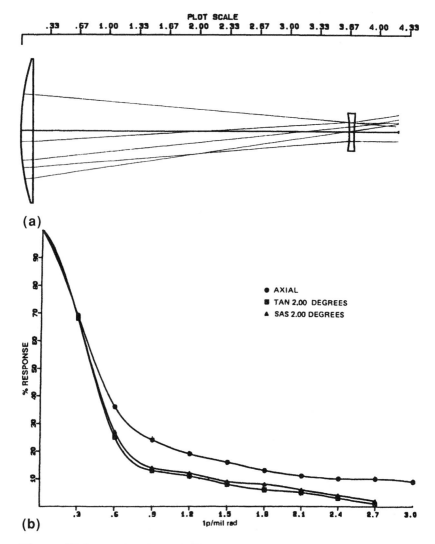

Figure 13-4 (a) 4× telescope. (b) Lens MTF.

Table 13-4 4× Telescope

	Radius	Thickness	Material	Diameter
0	0.00000	1.0000E+10	Air	
1	2.39778	0.1360	Plexiglas	1.440
2	0.00000	3.5228	Air	1.440
3	−0.96796	0.0500	Plexiglas	0.380
4	1.65724		Air	0.380

Note from the lens MTF plot that there is a rapid drop in response at 0.6 lp/milradian. This is a chromatic aberration contribution to the loss in contrast.

In figure 13-5 is shown an afocal power changer for use in a microscope. It is designed to rotate on a turret. A doublet is used with the usual microscope objective so that collimated light emerges. This goes to the power changer and thence to another doublet, which forms an image of the specimen to be viewed with an eyepiece. The power changer lenses are arranged so that light may pass through the two lenses (unit power); the lenses rotate to the position shown (0.75×); or the lenses rotate 180 degrees (1.333×). Following is lens data for Fig. 13-5.

Table 13-5 Power Changer

	Radius	Thickness	Material	Diameter
0	0.00000	1.00000E+10	Air	
1	−16.18482	0.1422	LF7	0.680
2	−0.84948	0.0504	SK16	0.680
3	2.79513	0.7857	Air	0.680
4	−7.39751	0.1167	LF7	0.900
5	1.33365	0.2242	SK16	1.040
6	−2.25821		Air	1.040

The overall length of the lens system is 1.319. The entrance pupil is in contact with the first lens surface.

In Figure 13-6 is shown an Albada viewfinder. This is a simple viewfinder used on some inexpensive cameras and on certain types of photometers. Following is lens data for Fig. 13-6. The overall length of the lens system is 1.148.

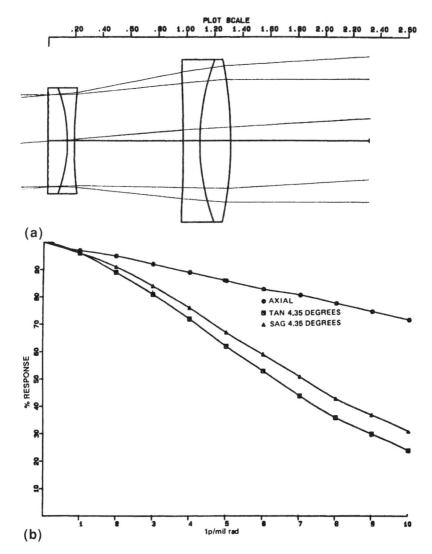

Figure 13-5 (a) Power changer 0.75 ×. (b) Lens MTF.

Afocal Optical Systems

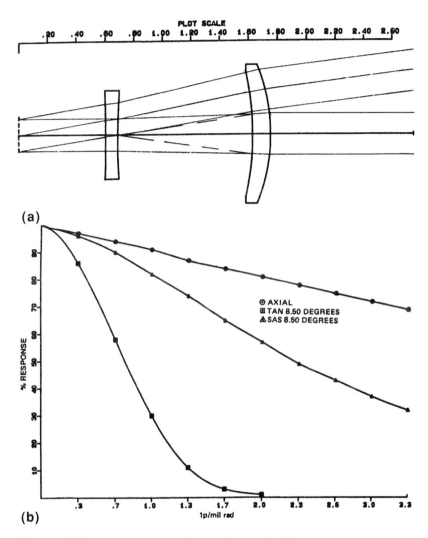

Figure 13-6 (a) Albada viewfinder. (b) Lens MTF.

Table 13-6 Albada Viewfinder

	Radius	Thickness	Material	Diameter
0	Ent. pupil	0.6000	Air	0.200
1	−5.80200	0.0820	K5	0.550
2	3.30100	0.9432	Air	0.550
3	−1.59700	0.1230	K5	0.780
4	−1.03200		Air	0.860

In place of the usual antireflection film, surface 3 ($R=-1.597$) is coated with a quarter-wave thickness of ZnS. This has a reflectivity of approximately 30%. On surface 2 there is evaporated with chrome (or Nichrome) the pattern to be projected. This viewfinder acts as a Galilean telescope of 1.29 power. Since the chrome pattern is at the focus (as shown by the dashed lines in Fig. 13-6) of the optical system consisting of the first lens and the mirror of surface 3 (EFL = 0.767), this pattern also appears as a bright image at infinity. The MTF plot is for the view through the system.

For security reasons, it is desirable for a homeowner to be able to look through a small hole in the door and see a wide field of view. It is obvious that a simple hole in the door provides a narrow field of view. However, a reversed Galilean type of telescope can provide the viewer with a wide field of view through a small door opening. Such door viewers are sold in hardware stores and are a common door accessory. Figure 13-7 shows such a door viewer. It has a 150 degree field of

Table 13-7 Door Scope

	Radius	Thickness	Material	Diameter
0	0.00000	1.00000E+07	Air	
1	0.99393	0.0600	K5	0.900
2	0.52338	0.1217	Air	0.620
3	0.00000	0.0400	K5	0.700
4	0.24502	0.1675	Air	0.360
5	−0.24502	0.0400	K5	0.360
6	0.00000	0.6016	Air	0.700
7	0.95167	0.1500	K5	0.480
8	−0.95167	0.5000	Air	0.480
9	Exit pupil		Air	0.196

Afocal Optical Systems 167

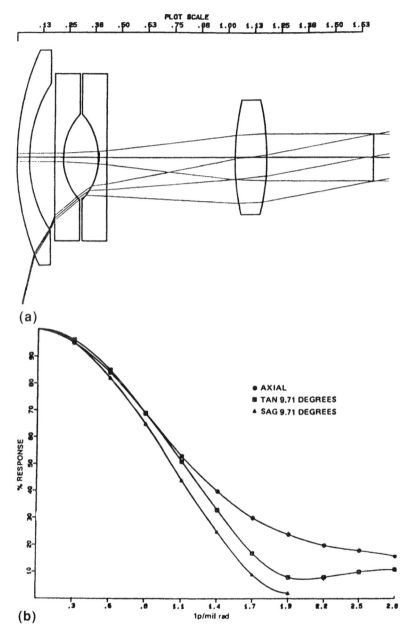

Figure 13-7 (a) Door scope. (b) Lens MTF.

view and a 5 mm exit pupil, 0.5 inch from the last lens surface. Of course, if the observer moves his eye sideways, the field of view is thereby increased. Following is lens data for Fig. 13-7.

The distance from the first surface to the last is 1.181. The entrance pupil is of diameter 0.04 and is located 0.399 from the first lens surface.

This system, like all wide-angle optical systems, exhibits considerable shift of the entrance pupil toward the front of the lens with field angle. The following table gives these shifts as measured from the paraxial pupil location.

Angle	Shift	Compression
37.5 deg	0.046	0.761
52.5	0.093	0.540
75	0.196	0.201

Also see Chap. 9 and design 17-4 for a 10× beam expander at 10.6 μm.

REFERENCES

Fraser, R. and McGrath, J. (1973). Folding camera viewfinder, U.S. Patent #3710697.
Itzuka, Y. (1982). Inverted Galilean finder, U.S. Patent #4348090.
Neil, I. A. (1983). Collimation lens system, U.S. Patent #4398786.
Neil, I. A. (1983). A focal refractor telescopes, U.S. Patent #4397520.
Rogers, P. J. (1983). Infra-red optical systems, U.S. Patent #4383727.
Wetherell, W. B. (1987). A focal Lenses, Chapter 3 in Volume 10 of *Applied Optics and Optical Engineering,* R. Shannon and J. Wyant, eds., Academic Press, New York.
Yanagimachi, M. (1979). Door scope, U.S. Patent #4172636.

14
Relay Systems

Relays are used to relay an image from one place to another. They are used in rifle sights, periscope systems, microscopes, and military infrared imaging systems. (This type of lens is also used for photocopy work.) Generally a field lens is required at an image plane to image the exit pupil of the previous group into the entrance pupil of the next group. The designer should keep in mind that a computer program, if allowed to vary the refractive index of this field lens, will select a glass of high index and low dispersion. Very little is gained by using glasses of refractive index greater than that of SK5. Also, a material must be selected with a low bubble code. Since this field lens is at an intermediate image, dirt, scratches, and other imperfections are imaged onto the final image. It is thus advisable to move the field lens away slightly from the intermediate image.

One-to-one imaging is desired since a completely symmetric lens may be used. This means that all transverse aberrations (distortion, coma, and lateral color) are zero.

Figure 14.1 shows a symmetric one-to-one relay (without its field lens) for a 1.2 diameter object. The lens focal length is 5.273.

The numeric aperture is

$$NA = \frac{0.5}{f^{\#}(M+1)} = 0.1$$

The lens focal length is 26.682. The KZFS4 elements help to reduce secondary color.

169

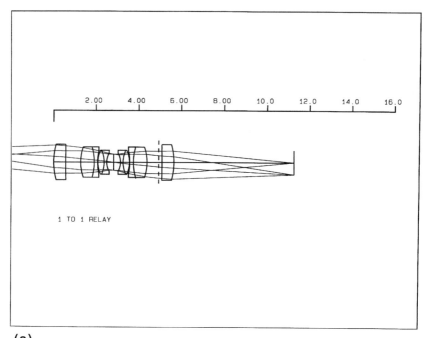

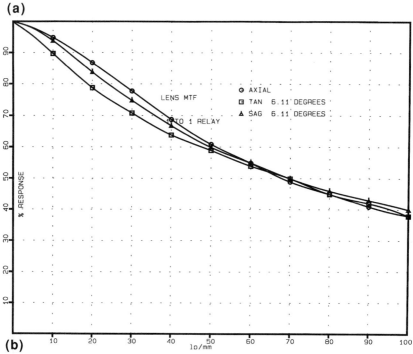

Figure 14-1 (a) One-to-one relay. (b) Lens MTF.

Table 14-1 1:1 Relay

	Radius	Thickness	Material	Diameter
0	0.00000	5.6018	Air	1.200
1	3.55526	0.5591	LAKN12	1.780
2	47.20269	0.7095	Air	1.780
3	2.46089	0.6425	LAKN12	1.520
4	−2.86878	0.1675	BAF4	1.520
5	2.82682	0.0138	Air	1.160
6	1.48812	0.2976	SSK4A	1.200
7	−1.52102	0.1209	KZFSN4	1.200
8	0.96535	0.3187	Air	0.900
9	Stop	0.3187	Air	0.812
10	−0.96535	0.1209	KZFSN4	0.900
11	1.52102	0.2976	SSK4A	1.200
12	−1.48812	0.0138	Air	1.200
13	−2.82682	0.1675	BAF4	1.160
14	2.86878	0.6425	LAKN12	1.520
15	−2.46089	0.7095	Air	1.520
16	−47.20269	0.5591	LAKN12	1.780
17	−3.55526	5.5973	Air	1.780

The $f^{\#} = 2.5$. The entrance pupil distance is 4.945. The distance from the first lens surface to the image is = 11.257.

Figure 14-2 shows a lens used in photocopy work. It images a 25 inch diameter object at unit magnification. Although this particular design is in the visual region, some of these lenses are designed to be used with orthochromatic film and so are designed for a shorter wavelength region. A general procedure in the graphic arts and printing trades is to create the various articles and pictures as a pasteup. This is then photographed to make the actual printing plate.

This design is unusual in that the MTF at full field is better than at mid-field. This is due to large sagittal errors at mid-field.

This lens is f/11 and has a focal length of 26.682. Copy lenses are generally fitted with an iris.

Figure 14-3 shows a copy lens for 0.6× magnification. It was designed to image a 5 inch diameter object into a 3 inch diameter image. It is f/3.5 (NA = 0.08928). Following is lens data for Fig. 14-3.

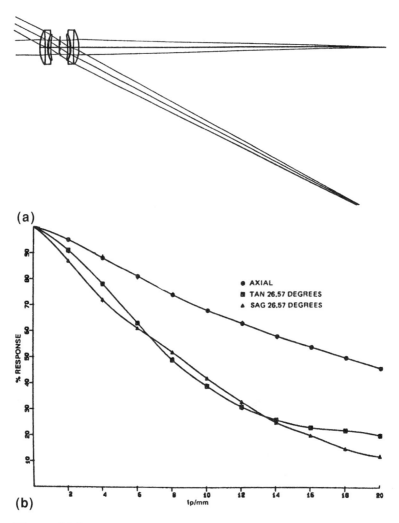

Figure 14-2 (a) Unit power copy lens f/11 675 mm F1. (b) Lens MTF.

Relay Systems

Table 14-2 Unit Power Copy Leans

	Radius	Thickness	Material	Diameter
0	0.00000	50.000	Air	50.00
1	4.64461	0.9205	SK14	4.960
2	−31.49862	0.3707	BALF5	4.960
3	3.56406	0.1643	Air	3.730
4	4.88823	0.3725	FK51	3.760
5	6.96846	1.4193	Air	3.500
6	Iris	1.4193	Air	2.040
7	−6.96846	0.3725	FK51	3.500
8	−4.88823	0.1643	Air	3.760
9	−3.56406	0.3707	BALF5	3.730
10	31.49862	0.9205	SK14	4.960
11	−4.64461	49.9560	Air	4.960

The distance from the first lens surface to the image is 56.451. The entrance pupil from the first lens surface is 3.363.

Table 14-3 0.6× Copy Lens

	Radius	Thickness	Material	Diameter
0	0.00000	16.6667	Air	5.000
1	2.20687	0.7634	LAKN12	2.380
2	−11.63771	0.0444	Air	2.380
3	−8.36718	0.3558	LF5	2.040
4	1.53422	0.1418	Air	1.680
5	3.02110	0.4607	LAKN7	1.740
6	6.51790	0.2662	Air	1.540
7	Stop	0.2441	Air	1.410
8	−2.69626	0.4767	LAKN7	1.500
9	−2.00098	0.1345	Air	1.740
10	−1.51146	0.1966	LF5	1.680
11	−11.30521	0.0160	Air	1.960
12	−25.44453	0.5831	LAKN12	2.150
13	−2.15957	9.1020	Air	2.150

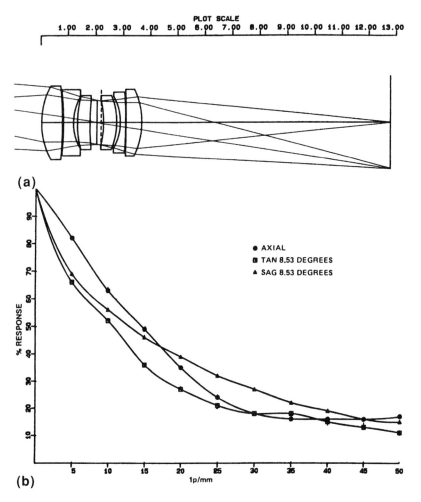

Figure 14-3 (a) 0.6× copy lens. (b) Lens MTF.

The entrance pupil distance from the first lens surface is 2.180. The distance from the first lens surface to the image is 12.785. The lens focal length is 6.944.

These elements tend to be rather thick. This helps reduce the Petzval sum. Spherical aberration is the main problem here. Distortion is 0.4%.

Relay Systems

Whenever a lens is to be fitted with an iris, be sure to allow at least 0.12 on both sides of the stop location for this iris mechanism.

Figure 14-4 shows a small three-element relay being used to create an erect image in a rifle sight. In a rifle sight, unlike a binocular, a long in-line optical system is desired. This facilitates mounting on the rifle. The front objective forms an image 0.277 behind the plano surface of the field lens. The relay then images this 1.552 from the last surface of the relay (1.064 in front of the eyepiece). At either of these intermediate image locations, a cross-line is placed. It may take the form of an actual wire stretched over a frame or a photographic pattern on thin glass. This

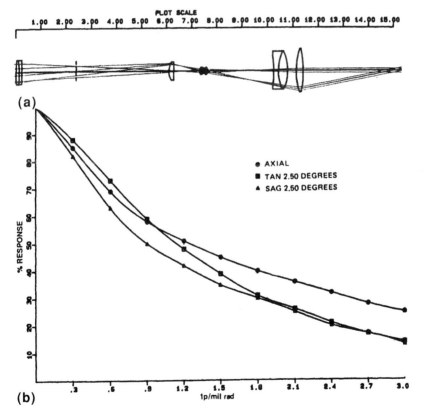

Figure 14-4 (a) Rifle sight. (b) Lens MTF.

cross-line may be adjusted transversely to bore sight the scope to the rifle. The eyepiece then forms a virtual image of this to the eye.

Note that on the lens MTF plot, the abscissa (as in all the afocal systems discussed here) has units of lp/milradian. This is sometimes called cycles/milradian. 3 lp/milradian is approximately 1 minute of arc, the limit of visual acuity.

Due to rifle recoil, a very large eye relief is required (3.92 in this case). To obtain this long eye relief, the entrance pupil is positioned substantially inside the front objective. The field lens causes the chief ray to cross the axis very close to the last surface of the relay lens. The image of this by the eyepiece is the exit pupil. The magnification of the system is 4; the field of view = 5 degrees. The entrance pupil diameter is 16 mm. There is 20% vignetting at full field.

Since a rifle sight is often exposed to rain, fog, snow, etc., it is important that glass types used for front and rear lens surfaces be selected for high weather and stain resistance. In the usual glass catalogs, the most resistant materials are indicated with low numbers

Table 14-4 Rifle Sight

	Radius	Thickness	Material	Diameter
0	0.00000	1.00000E+07	Air	
1	2.69406	0.1417	BK7	0.870
2	−2.61456	0.0984	F2	0.870
3	217.12603	5.8379	Air	0.870
4	0.58062	0.1968	BK7	0.650
5	0.00000	1.0416	Air	0.650
6	0.23026	0.0915	SSK4A	0.260
7	2.23574	0.0530	Air	0.260
8	−0.26421	0.0322	SF2	0.210
9	0.18961	0.0396	Air	0.210
10	0.44149	0.0942	SSK4A	0.250
11	−0.22138	2.6157	Air	0.250
12	−3.19380	0.1783	SF4	1.100
13	1.49598	0.3679	SK2	1.360
14	−1.49598	0.3356	Air	1.360
15	2.24645	0.2754	BAF3	1.540
16	−3.51257	3.9275	Air	1.540

Relay Systems

like 0, 1, or 2, whereas glasses with poor stain and weather resistance have codes of 4 or 5. In some of the glass catalogs, weather resistance is termed climatic resistance and is a measure of the effect of water vapor in the air on the glass. Staining is a result of the glass being in contact with slightly acidic water (often due to a fingerprint on the glass). (Figure 35-4 shows a zoom rifle sight that might be of interest.)

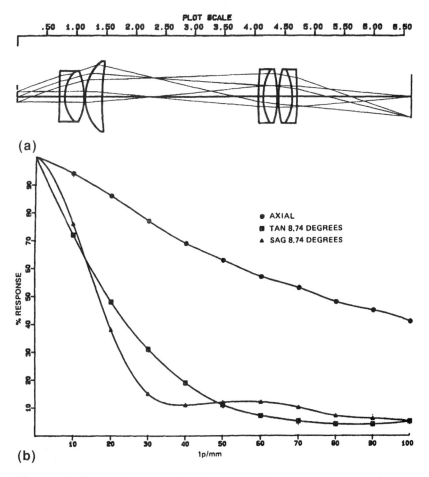

Figure 14-5 (a) Eyepiece relay. (b) Lens MTF.

Table 14-5 Eyepiece Relay

	Radius	Thickness	Material	Diameter
0	0.00000	1.00000E+07	Air	
1	15.49980	0.0880	SF8	0.760
2	0.49076	0.3281	BK7	0.760
3	−0.84226	0.0182	Air	0.760
4	0.65510	0.2806	K5	1.060
5	2.99551	2.6373	Air	1.000
6	3.80120	0.0808	BASF2	0.800
7	0.75419	0.2317	SK5	0.800
8	−1.39190	0.0201	Air	0.800
9	1.39190	0.2317	SK5	0.800
10	−0.75419	0.0808	BASF2	0.800
11	−3.80120	1.9231	Air	0.800

The distance from the first lens surface to the image is 5.920. Distortion = 1.9%.

The distance from the first surface to the last is 11.400. Distortion = 0.9%. At the edge of the field there is some primary lateral color. For manufacturing reasons, the rear surface of the front objective (R= 217.126) should be set plane. The entrance pupil is 2.383 from the first lens surface (0.630 diameter).

Figure 14-5 shows an eyepiece relay similar to the above rifle sight. It was designed to view a 16 mm diameter object with a 4 mm diameter entrance pupil. The entrance pupil distance from the first lens surface is 0.717. This was intended as a camera viewfinder for 16 mm cinematography. The eye relief is 18 mm.

Note that this design features a relay composed of two identical cemented doublets. Although this simple construction is economical to manufacture, it has the usual astigmatism common to cemented doublets. Due to this image degradation at the edge of the field, the MTF data is for half field.

Figure 14-6 shows a high numeric aperture (0.4) relay. The object diameter is 0.32 and is imaged at a 1/5 reduction. The entrance pupil is 3.571 from the first lens surface. The object to-image distance is 8.487. Distortion is 0.5% (pincushion).

Relay Systems

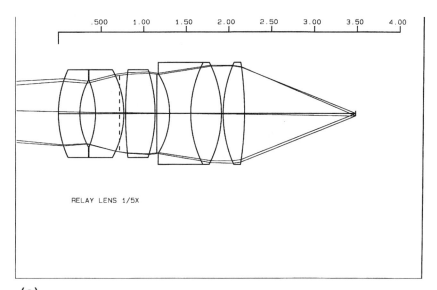

(a)

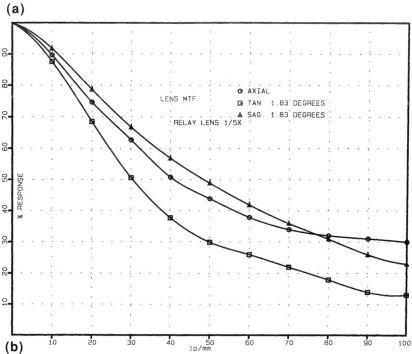

(b)

Figure 14-6 (a) 1/5× relay lens. (b) Lens MTF.

Table 14-6 1/5× Relay

	Radius	Thickness	Material	Diameter
0	0.00000	5.0000	Air	0.320
1	1.41762	0.2500	SF4	1.100
2	0.91371	0.1829	Air	0.840
3	−1.19430	0.3304	LAFN21	0.840
4	−1.20238	0.0209	Air	1.100
5	5.54790	0.3500	LAFN21	1.100
6	−1.89990	0.0205	Air	1.100
7	Stop	0.1508	Air	0.988
8	−0.99151	0.2493	SF4	1.000
9	1.48728	0.3602	LAK10	1.280
10	−1.50369	0.0207	Air	1.280
11	1.67360	0.2523	LAFN21	1.280
12	−4.49468	1.2986	Air	1.280

REFERENCES

Cook, G. (1952). Four component optical objective, U.S. Patent #2600207.
Itoh, T. (1982). Variable power copying lens, U.S. Patent #4359269.
Kawakami, T. (1976). Symmetrical type four component objective, US Patent #3941457.
Shade, W. E. (1967). Projection printer lens, U.S. Patent #3320017.
Terasawa, H. (1985). Projection lens, U.S. Patent #4560243.
Tibbetts, R. E. (1971). High speed document lens, U.S. Patent #3575495.
Yonekubo, K. (1982). Afocal relay lens system, U.S. Patent #4353624.

15
Catadioptric and Mirror Optical Systems

A mirror has substantially less spherical aberration than a lens of equivalent focal length. In addition, the spherical aberration of a concave mirror is opposite in sign to that of a positive lens. Therefore several people (Maksutov, 1944; Bouwers, 1950) have proposed the use of a weak negative lens in conjunction with a concave mirror. Schmidt (Hodges, 1953) proposed a weak aspheric corrector at the center of curvature of a spherical mirror.

In the Cassegrain system, two mirrors are required. The primary has a hole in it and the smaller secondary is attached to (or is actually a part of) one of the front corrector lenses. Due to vignetting, fields of view are generally limited to about 15 degrees.

The computer program must have the following features;

1. Surface radii, material, and thickness must be capable of being tied or set equal to another surface.
2. Axial distances from one surface to another must be capable of being bounded.
3. Beam diameter at the secondary mirror has to be limited.
4. Delete tracing of the chief ray.

Referring to Fig. 15-1 for a Cassegrain mirror system showing a primary and secondary mirror separated by a distance T, forming an image a distance B from the secondary, and assuming a sign convention in which T and B are positive and R_p and R_s are negative as shown, then

$$F = \frac{R_p R_s}{2R_p - 2R_s + 4T} = \text{the effective focal length}$$

Let H be the height ratio of a paraxial axial ray at the secondary mirror divided by the height at the primary. If the object is at infinity, then

$$H = \frac{R_p - 2T}{R_p} \quad \text{and} \quad B = \frac{2R_s T + R_s R_p}{4T - 2R_s + 2R_p} = HF$$

Most practical systems require that B should be at least as large as T (the image is outside the primary mirror). Also the obscuration by the secondary needs to be minimized. Assigning a reasonable value for H of 0.3 and for a focal length (F) of 100, we obtain the data in Table 15-1 (B is obviously 30 for all values in this table).

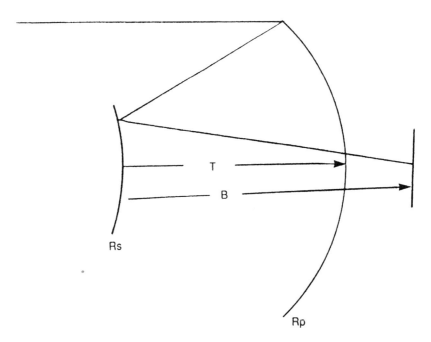

Figure 15-1 Cassegrain mirror system.

Table 15-1 Primary and Secondary Mirror Geometry, F=100

R_p	R_s	T
−171.428	−360.000	60.000
−122.449	−94.737	42.857
−95.238	−54.545	33.333
−77.922	−38.298	27.273
−65.934	−29.508	23.077
−57.143	−24.000	20.000
−50.420	−20.225	17.647
−45.113	−17.476	15.789
−40.816	−15.385	14.286
−37.267	−13.740	13.043

See the program Mirror in Appendix E. If only two mirrors are used then for a flat field (zero Petzval sum), $R_p = R_s = -140$ with $T = 49$. However, to correct for other aberrations, it is necessary to add some refractive elements as shown in the following examples. The above table is useful in obtaining a starting solution.

The effect of obscuration must be considered in evaluating the image. Diffraction, as a result of the obscuration, increases the energy in the outer portions of the image at the expense of the central bright spot. This reduces the MTF response at low spatial frequencies and increases the response at higher spatial frequencies (Mahajan, 1977).

Figure 15-2 shows a 15 inch focal length f/3.333 lens that covers the spectral region from 3.2 to 4.2 microns. It has a 2.3 degree FOV.

Table 15-2 Cassegrain Lens

	Radius	Thickness	Material	Diameter
0	0.00000	1.00000E+08	Air	
1	43.46138	0.3797	ZnS	4.600
2	0.00000	4.0000	Air	4.600
3	−9.79524	0.5500	ZnS	4.160
4	−11.90806	−0.5500	Mirror/ZnS	4.380
5	−9.79524	−3.8200	Air	4.160
6	−6.38180	3.9792	Mirror	1.450

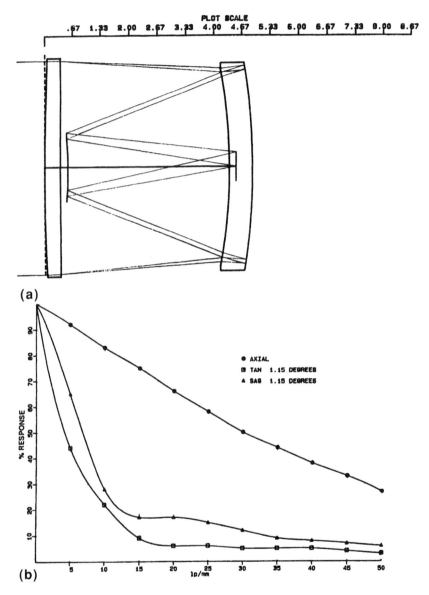

Figure 15-2 (a) Cassegrain lens, ZnS, 3.2–4.2 μm. (b) Lens MTF.

The distance from the first lens surface to the image is 4.539. Distortion is 0.7% (pincushion). The entrance pupil is in contact with the first lens surface.

As an alternate, this design could have used zinc selenide (ZnSe) in place of the zinc sulfide (ZnS). ZnSe has a slightly higher refractive index, has a larger V value, and is transparent over a larger spectral region.

You will note that for the system as shown, a ray may pass through the front corrector and go directly to the image. To prevent this from occurring, baffles are required. This is accomplished with a lens extension tube placed in front of the lens plus a tube in the hole of the primary mirror as well as a shield at the secondary mirror.

Placing baffles to prevent these unwanted rays is a complex task. Although there are computer programs available to help the designer in this specific task, a simple method is to make a large and careful layout showing the limiting rays. Then trace rays that would penetrate the system without making the necessary reflections at primary and secondary mirrors. From this layout, the required baffles may be determined.

Also note that the front lens is positive. However, in conjunction with the Mangin primary mirror, the refractive elements have negative power.

Figure 15-3 shows a 4 inch focal length f/1.57 Cassegrain with a 15 degree field of view for the visual region. The front corrector is of the usual negative form. Note that the secondary mirror has the same radius as the front refracting surface. This is a manufacturing convenience, since with a mask, a reflecting coating is first applied to the inside diameter (corresponding to the secondary mirror), an antireflection coating is applied to the entire surface, and then the mirror surface is protected with a black paint.

Due to its compact form and low f#, a variation of this design was extensively used during the Korean war as a night vision device. Excellent imagery can be obtained to wavelengths as long as 0.85 μm With an image intensifier tube, and an eyepiece it was fitted to a rifle to allow a soldier to function at night with the aid only of starlight.

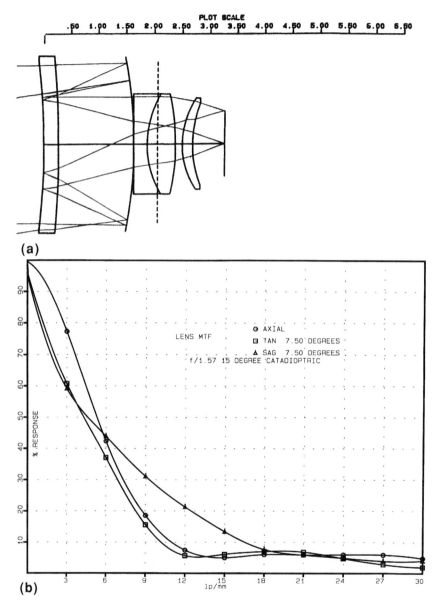

Figure 15-3 (a) Starlight scope objective, $f/1.57$. (b) Lens MTF.

Table 15-3 Starlight Scope Objective

	Radius	Thickness	Material	Diameter
0	0.00000	1.00000E+07	Air	
1	−9.69598	0.2750	BK7	2.750
2	−12.89434	1.3662	Air	2.850
3	−6.50950	−1.3662	Mirror	2.800
4	−12.89434	−0.2750	BK7	2.850
5	−9.69598	0.2750	Mirror/BK7	2.750
6	−12.89434	1.3662	Air	2.850
7	−40.57582	0.2311	SF2	1.620
8	1.46999	0.5096	LAKN7	1.620
9	−3.22355	0.1192	Air	1.620
10	1.25132	0.1832	SK16	1.480
11	1.68256	0.5849	Air	1.340

The entrance pupil of 2.54 diameter is located 2.056 from the first surface. The distance from the first lens surface to the image of 3.269. At 7.5 degrees off-axis the distortion is 0.6% (pincushion). There is some vignetting (the tangential component of the entrance pupil is 0.85 of axial. Relative illumination is 0.8.

Figure 15-4 shows a 1000 mm focal length f/11 Cassegrain for use with a 35 mm single lens reflex camera (visual region 43 mm diagonal image). In order to reduce the diameter of the secondary mirror it was made the stop. Distortion is less than 0.1%. The entrance pupil of 3.58 diameter is located 34.217 from the first surface.

Table 15-4 1000mm Focal Length Cassegrain

	Radius	Thickness	Material	Diameter
0	0.00000	1.00000E+07	Air	
1	−5.16251	0.4525	BK7	5.050
2	−5.41781	8.5519	Air	5.280
3	−22.42740	−8.1591	Mirror	4.920
4	−9.88275	7.9423	Mirror	1.080
5	−4.02610	0.1388	BK7	1.400
6	21.05001	0.0180	Air	1.500
7	3.38011	0.3947	SF1	1.500
8	4.09688	1.8171	Air	1.400

188   *Chapter 15*

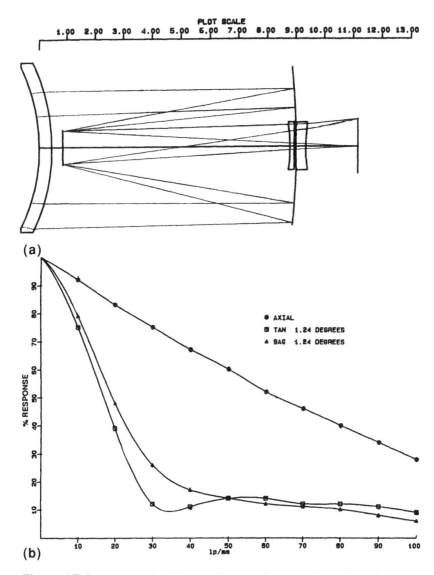

Figure 15-4   100-mm focal length Cassegrain lens. (b) Lens MTF.

The distance from the first lens surface to the image is 11.156. Although the system as shown has no vignetting, there may be a small amount after one introduces the tubular baffle required around the rear lenses inside the primary mirror.

Figure 15-5 shows a Cassegrain telescope objective with a 50 inch focal length. It is f/14 with a 2 degree FOV. This is a modification of a very popular commercially available telescope objective for SLR photography and amateur astronomy.

Table 15-5 50 Inch Focal Length Telescope Objective

	Radius	Thickness	Material	Diameter
0	0.00000	1.00000E+08	Air	
1	−13.69544*	0.6000	Quartz	5.250
2	−13.88461	10.9993	Air	5.400
3	−32.61588	−10.9993	Mirror	5.020
4	−13.88461	−0.6000	Quartz	1.220
5	−13.69544*	0.6000	Mirror/Quartz	1.080
6	−13.88461	10.4992	Air	1.100
7	10.01837	0.1926	BK7	1.700
8	21.85435	1.0287	Air	1.700
9	−7.29090	0.1926	SF1	1.700
10	−9.45275	3.5358	Air	1.700

The distance from the first lens surface to the image is 16.049. The entrance pupil of 3.57 diameter is located 48.133 from the first surface. Surface 5 acts as a stop. Lines marked with an asterisk denote aspheric surfaces with the following relationship:

$$X = \frac{0.0730172 Y^2}{1 + \sqrt{1 - 0.0057659 Y^2}} - 1.02289 \times 10^{-5} Y^4 + 3.91550 \times 10^{-8} Y^6$$

This represents a surface with a slight deviation from an ellipse with axis normal to the optical axis. Distortion is 0.4% (pincushion).

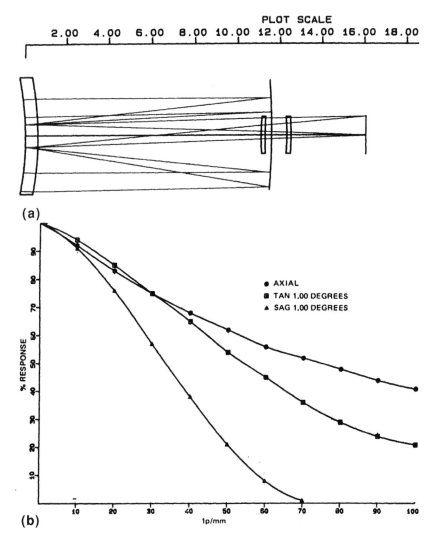

Figure 15-5 (a) 50-in. focal length telescope objective. (b) Lens MTF.

Figure 15-6 shows a 10 inch focal length f/1.23 Cassegrain. It was designed for a 40 mm diameter image tube (FOV = 9 degrees). The design is remarkable in that only one material is used; it nevertheless has negligible chromatic aberration. Although designed for the visual region, it is useable over an extended wavelength region (see Shenker, 1966).

Table 15-6 10 Inch Focal Length Cassegrain

	Radius	Thickness	Material	Diameter
0	0.00000	1.00000E+08	Air	
1	19.68695	0.5412	BK7	8.250
2	0.00000	2.1942	Air	8.250
3	−13.00717	0.3509	BK7	7.800
4	−18.93058	1.8118	Air	7.900
5	−10.38543	0.3987	BK7	7.700
6	−15.04273	3.7922	Air	7.900
7	−12.97655	−3.7922	Mirror	8.000
8	−15.04273	3.2927	Mirror	3.660
9	−24.65875	0.1590	BK7	2.160
10	−10.22278	0.0200	Air	2.160
11	2.18312	0.3643	BK7	1.980
12	2.28226	0.3555	Air	1.700

The distance from the first lens surface to the image is 9.488. The entrance pupil of 8.13 diameter is located 0.225 in front of the first lens surface. This serves as a stop for the lens. Baffling may be accomplished with a cone surrounding the two rear lenses and a short lens hood. There is some vignetting. Distortion is less than 0.1%.

The tangential pupil size in relation to the axial dimension is as in the table.

Angle (°)	Tangential component
2.25	0.92
4.50	0.72

In 1930 Bernard Schmidt completed his famous camera. At the center of curvature of a spherical mirror, he placed an aspheric plate

192 Chapter 15

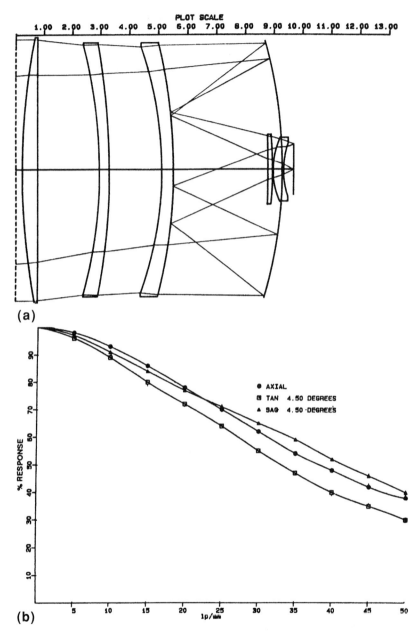

Figure 15-6 (a) 10-inch FL *f*/1.23 Cassegrain lens. (b) Lens MTF.

with the flat side facing the mirror. It was a concentric system with the image being formed halfway between the corrector and the mirror. The image surface was spherical and of radius equal to the focal length.

Schmidt manufactured his corrector plate by calculating the elastic deformation of a plate held by a ring of known diameter. By means of a vacuum pump, he then deformed this plate the proper amount, ground one surface flat, and then released the vacuum, allowing the plate to assume its proper aspheric shape.

The Palomar Observatory has two Schmidt cameras, one with an 18 inch and the other with a 48 inch aperture. These are used to map the entire night sky of the northern hemisphere.

Figure 15-7 shows a Schmidt objective. It is f/1.8 and covers a field of 7 degrees. The focal length is 10.

Table 15-7 Schmidt Objective

	Radius	Thickness	Material	Diameter
0	0.00000	1.00000E+08	Air	
1	0.00000	0.3000	Quartz	5.600
2	0.00000	20.0000	Air	5.600
3	−20.00000	−9.9861	Mirror	8.100

The image surface is spherical of −10 radius. The aperture stop is in contact with the second surface. An entrance pupil of diameter 5.556 is located 0.205 from the first surface.

The equation for the first surface, (it has a paraxial curvature of 0.0) is

$$X = 1.024119 \times 10^{-5} Y_4 - 1.804486 \times 10^{-5} Y^6 + 1.831413 \times 10^{-6} Y^8 - 6.879236 \times 10^{-8} Y^{10}$$

The disadvantage of this system is the inconvenient position of the image inside the objective as well as its curvature. Its advantage lies in simplicity of construction and its ability to give excellent resolution over a large wavelength region and field of view. A Cassegrain version has been proposed (Baker, 1940). This requires an aspheric primary mirror with a spherical secondary. The radii are the same, so we have a zero Petzval sum with a flat field. However, the author's experience

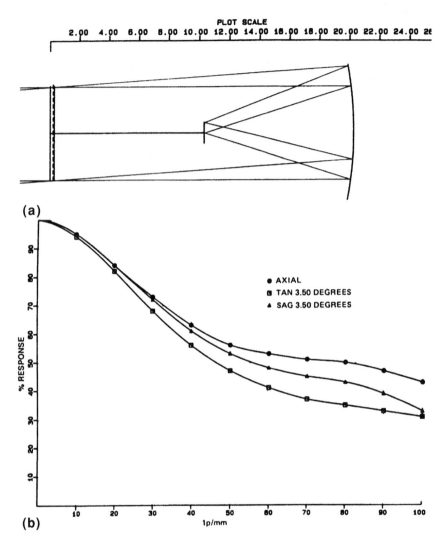

Figure 15-7 (a) Schmidt objective. (b) Lens MTF.

Catadioptric and Mirror Optical Systems

with this modification has been disappointing. I found it far better to use one of the all-spherical systems as given above.

All of the above systems (with perhaps the exception of the Schmidt system) are telephoto lenses. That is, the overall length of the lens is much less than the effective focal length. They also have relatively short (as a fraction of their focal length) back focal lengths. By reversing the role of primary and secondary mirrors, we may obtain an inverted telephoto type of design. Here the overall length will be substantially longer than the effective focal length, and the back focal length will be longer than the effective focal length. This is a useful system for lenses of relatively short focal length where a long working distance is required.

Such an inverted telephoto type of design is shown in Fig. 15-8. It is f/2.5 and has an image of 0.2 diameter. The entrance pupil is in contact with the first surface.

Table 15-8 Reflecting Objective

	Radius	Thickness	Material	Diameter
0	0.00000	1.00000E+07	Air	
1	1.89149	−1.7946	Mirror	0.800
2	3.63835	1.7946	Mirror	2.550
3	1.89149	0.3193	BK7	1.800
4	1.49667	0.3410	Air	1.600
5	4.41392	0.3000	SF1	1.520
6	4.53570	2.8493	Air	1.360

The distance from the large mirror surface ($R=3.6384$) to the image is 5.605. Distortion is negligible.

This lens has considerable inward curving field. The best off-axis image is obtained 30 µm inside the best axial image surface. MTF at the edge of the field would become at 100 lp /mm 33% tan and 15% sag.

The computer program must have the ability to bound the ray heights at the mirror surfaces. Since there is a hole in the large mirror, the first mirror surface must have sufficient curvature to cause the outer rim rays to strike the large mirror outside of this hole. This lens would make an ideal long working distance microscope objective (see Chap. 11).

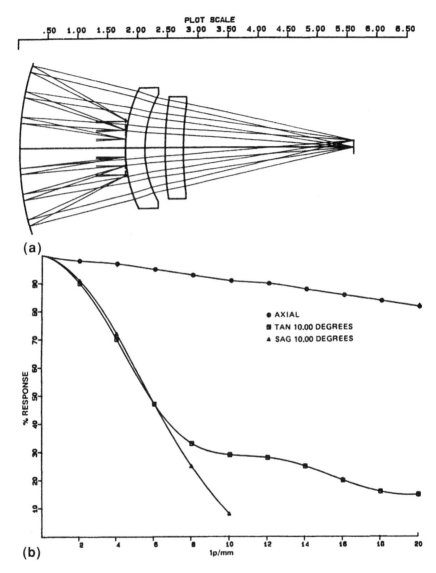

Figure 15-8 (a) Reflecting objective f/2.5. (b) Lens MTF.

Catadioptric and Mirror Optical Systems 197

In Fig. 15-9 is shown a long effective focal length (250 in.) f/10 objective. Image size is 1.703, so it is suitable for a 35 mm SLR camera. Both mirrors are hyperbolas, so this may be considered as a modified Ritchey–Chretien design (Rutten, 1988). The lens data is given in the following table. The entrance pupil is located 37.297 in front of the primary mirror (in line with the secondary mirror).

Table 15-9 250 Inch Cassegrain

	Radius	Thickness	Material	Diameter
0	0.00000	1.00000E+08	Air	
1	−101.99116	−37.2973	Mirror	25.350
2	−36.67501	47.4148	Mirror	7.100
3	18.08219	0.2500	SK16	2.170
4	10.42742	0.3600	LF5	2.170
5	51.26888	1.1030	Air	2.170
6	−9.45943	0.2500	LLF1	1.960
7	2.80216	0.5000	LF5	1.960
8	27.65025	5.6949	Air	1.960

The conic coefficients are (A_2 as given in the aspheric equation in Chap. 1)

−1.10533   primary mirror (surface 1)
−3.77539   secondary mirror (surface 2)

The distance from this mirror to the image is 55.572 in. Considerable baffling will be required as well as a spider mechanism to hold the secondary mirror in place. Diffraction effects for this are not considered in the MTF plot shown in 15-9b.

To illustrate the importance of the two cemented doublets in the rear, in Fig. 15-10 is shown a Ritchey–Chretien design with the same

Table 15-10 Ritchey–Chretien

Surf	Radius	Thickness	Material	Diameter
0	0.0000	1.0000E+11	Air	
1	Stop	63.0701	Air	25.000
2	−187.5040	−63.0701	Mirror	25.427
3	−98.1833	81.8111	Mirror	8.773

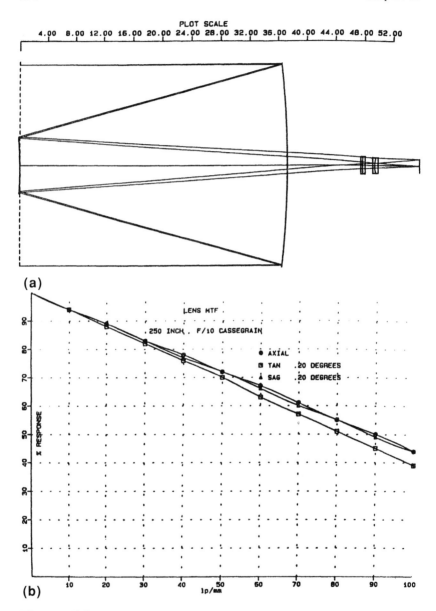

Figure 15-9 (a) 250-inch, $f/10$ Cassegrain lens. (b) Lens MTF.

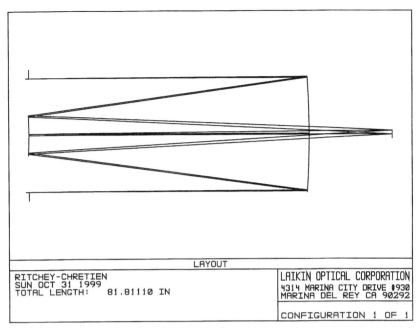

(a)

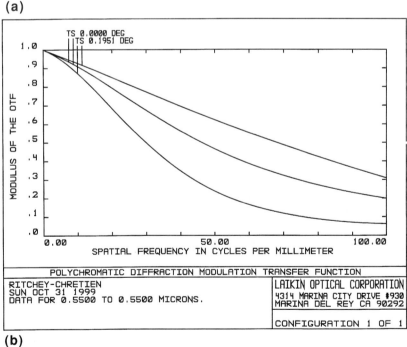

(b)

Figure 15-10 Ritchey–Chretien design. (b) Lens MTF.

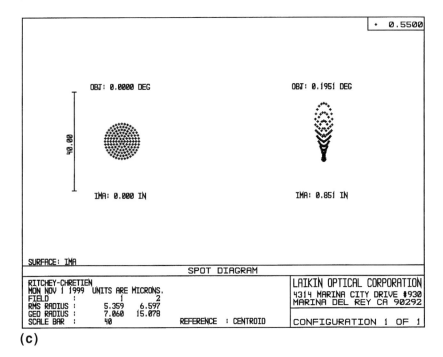

(c)

Figure 15-10 Continued

focal length, f#, and field of view as the above design. However, it consists of only two hyperbolic mirror (Rutten, 1988).

The conic coefficients are

−1.1184   primary mirror (surface 2)
−6.3236   secondary mirror (surface 3)

Note that the design of 15-9 has less obscuration by the secondary mirror. The two doublets in the rear help to correct the off-axis aberrations: coma, astigmatism, and field curvature. In Fig. 15-10c is shown the spot diagram for the system of Fig. 15-10a. Note the effect of coma and astigmatism on the spot diagram.

REFERENCES

Abel, I. R. (1983). Compact optical system, U.S. Patent #4411499.
Amon, M. Rosen, S., and Jackson, B. (1971). Large objective for night vision, *Applied Optics,* 10:490.

Amon, M. (1973). Large catadioptric objective, U.S. Patent #3711184.
Baker, J. G. (1940). A family of flat field cameras, equivalent in performance to the Schmidt camera, *Pro. Am. Phil. Soc., 82:*339.
Barnes, W. J. (1979). Optical Materials—Reflective, Chapter 4, Volume 7, *Applied Optics and Optical Engineering* R. S. Shannon and J. Wyant, (eds.), Academic Press, New York.
Blakley, R. (1995). Modification of the classic Schmidt telescope, *Optical Engineering, 34:*1471.
Blakley, R. (1996). Cesarian telescope optical system, *Optical Engineering, 35:*3338.
Bouwers, A. (1950). *Achievements in Optics,* Elsevier, New York.
Bowen, I. S. (1960). *Schmidt Cameras, Telescopes* (Kuiper, G. P. and Middlehurst, B. M., eds.), University of Chicago Press, Chicago, Chap. 4.
Bruggemann, H. P. (1968). *Conic Mirrors,* Focal Press, New York.
Hodges, P. C. (1953). Bernard Schmidt and his reflector camera (Ingalls, A., ed.), *Amateur Telescope Making, 3:*365, Scientific American.
Korsch, D. (1991). *Reflective Optics,* Academic Press, New York.
Kuiper, G. P. and Middlehurst, B. M. (1960). *Telescopes,* University of Chicago Press, Chicago.
Lucy, F. A. (1941). Exact and approximate computation of Schmidt cameras, *JOSA, 31:*358.
Lurie, R. (1975). Anastigmatic catadioptric telescope, *JOSA, 65:*261.
Mahajan, V. N. (1977). Imaging with obscured pupils, *Optics Letters, 1:*128.
Maksutov, D. D. (1944). New catadioptric meniscus system, *JOSA, 34:*270.
Maxwell, J. (1971). *Catadioptric Imaging Systems,* Elsevier, New York.
Powell, J. (1988). Design of a 300 mm focal length f/3.6 spectrographic objective, *Optical Engineering, 27:*1042.
Puryayev, D. T. and Gontcharov, A. V. (1998). Aplanatic four-mirror system for optical telescopes, *Optical Engineering, 37:*2334.
Rayces, J. L. (1975). All spherical solid catadioptric system, U.S. Patent #3926505.
Rutten, H. and van Venrooij, M. (1988). *Telescope Optics,* Willmann-Bell, Richmond, VA.
Shimizu, Y. (1972). Catadioptric telephoto objective lens, U.S. Patent #3632190.
Shenker, M. (1966). High speed catadioptric objective, U.S. Patent #3252373.
Stephens, R. E. (1948). Reduction of sphero-chromatic aberration in catadioptric systems, *J. Res. NBS 40:*467.

16
Periscope Systems

The periscope system discussed here is not of the long submarine type but is rather an extended optical system used in special effects cinematography. Such an elongated lens system allows the cinematographer to get a very bulky camera into an otherwise inaccessible place. Some cinematographers have erroneously concluded that the depth of field is greater than that obtained from the usual depth of field tables given in cinematography manuels (Samuelson, 1998). One reason for the apparent discrepancy is that these depth of field tables are measured from the lens front principal plane (a nodal point). For periscope systems, this principal plane is usually well within the body of the periscope giving an apparently short near depth of focus distance. Another reason is that these systems generally have many lens surfaces and a large distance between lens groups. Even with modern high efficiency antireflection coatings and threads and sand-blast finish on the interior walls, there is still substantial veiling glare in the final image. This results in a loss of contrast, giving the appearance of enhanced depth of field. Such a device is shown in figure 16-1.

A mirror is placed at the front external entrance pupil. This allows the cinematographer to scan the field in a vertical direction. (This assumes that the periscope tube is vertical and so the axis of the front mirror is horizontal.) Behind the relay is a penta-roof prism which deviates the beam 90 degrees and yields an image of the correct orientation on the film. This replaces the normal camera lens. Lens effective focal length is 1 and yet the length from mirror pivot to film center is nearly 27 inches (this considers the fold created by the penta-roof prism). The length from the first lens surface to the image is 33.197.

Table 16-1 25mm Focal Length Periscope

	Radius	Thickness	Material	Diameter
0	0.00000	1.00000E+05	Air	
1	−1.74000	0.1309	LAKN9	1.340
2	−1.95700	0.1969	LAF2	1.600
3	−1.18700	0.9965	Air	1.600
4	0.00000	0.2238	LAKN7	2.320
5	−4.78000	0.8560	Air	2.320
6	2.90200	0.8068	SK4	2.600
7	−2.03200	0.2000	SF1	2.600
8	0.00000	15.5160	Air	2.600
9	2.54500	0.2602	LAF20	1.930
10	1.82000	0.6260	F6	1.930
11	0.00000	0.9772	Air	1.930
12	−7.96000	0.3254	FK51	1.240
13	−2.20600	0.3111	LLF1	1.450
14	0.00000	0.1871	Air	1.450
15	2.00100	0.4431	LAKN22	1.450
16	−1.32500	0.1416	SF5	1.450
17	0.91400	0.2715	Air	0.790
18	Iris	0.2715	Air	0.810
19	−0.65100	0.1461	SF5	0.830
20	2.63000	0.5657	LAKN22	1.450
21	−1.30800	0.0202	Air	1.450
22	−3.31300	0.2809	LAF2	1.450
23	−1.98600	0.0198	Air	1.930
24	−6.34700	0.5149	LAF20	1.620
25	−2.03200	0.0999	Air	1.930
26	0.00000	5.2561	SK16	1.710
27	0.00000	3.5520	Air	1.710

Lens $f\#$ is 4.5. Lens transmittance (considering the aluminized front mirror and the silvered surfaces on the penta-roof prism) is 0.658.

$$T^{\#} = \frac{f\#}{\sqrt{\text{transmittance}}} = 5.6$$

Distortion is 1.9%. The rear group of lenses moves forward to focus from distant objects to less than 2 inches in front of the front mirror. This mirror is located at the entrance pupil, which is 1.153 in front of

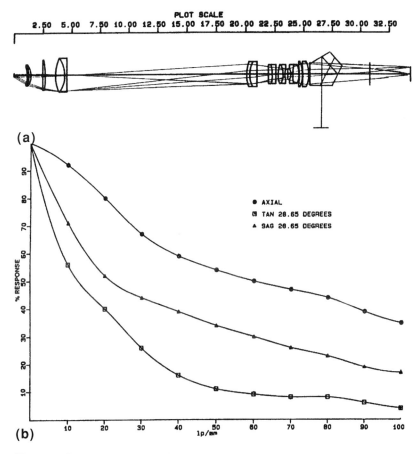

Figure 16-1 (a) 25-mm focal length periscope. (b) Lens MTF.

the first lens surface. This type of internal focusing is required when one realizes the large sizes of the periscope and camera. It would be difficult, mechanically, to move the periscope in relation to the camera. It is far simpler to move a lens group inside the periscope housing. This allows the periscope to be rigidly attached to the camera.

Determining this focusing movement is aided by being able to trace the system in reverse. That is, we move the relay group an increment forward and then trace (paraxial) from the image location to determine the object location. We then obtain the data in Table 16-2.

Table 16-2 Periscope Focusing (for 35 mm wide film)

Objd	T(9)	T(26)
infinity	15.516	0.100
7.74	15.316	0.300
4.00	15.016	0.600
3.07	14.716	0.900
2.90	14.616	1.000
2.56	14.316	1.300
2.37	14.016	1.600
2.24	13.716	1.900
2.16	13.416	2.200

The entire unit is fitted on a rotary mount such as to be able to scan 360 degrees in azimuth. These lenses are generally equipped with motors so that the iris, focus, front mirror, and rotary mount can be remotely controlled.

Figure 16-2 shows a periscope for use with 65 mm wide film. Referring to Appendix A, this film format has a diagonal of 2.101. This periscope has a focal length of 1.462, is f/8, and has a FOV of 71.4 deg.

The entire rear assembly, with the iris included, moves toward the film for focusing. Table 16-3 gives these movements.

This periscope differs from the previous in that there is no erecting/folding prism. The system is normally used in a vertical position. A mirror at the entrance pupil deviates the beam 90 deg. (As in the above case, this mirror moves under motor control.) Between the iris and the film is another mirror to deviate the beam again 90 deg. The camera is then oriented upside down so that an erect image is obtained on the film.

Note that the iris is actually external to the lens system. This causes considerable aberration of the entrance pupil. That is, the computer program must increase the diameter of the entrance pupil (in the tangential orientation) so that the diameter of the beam at the iris location is the same for the off-axis bundle as the axial.

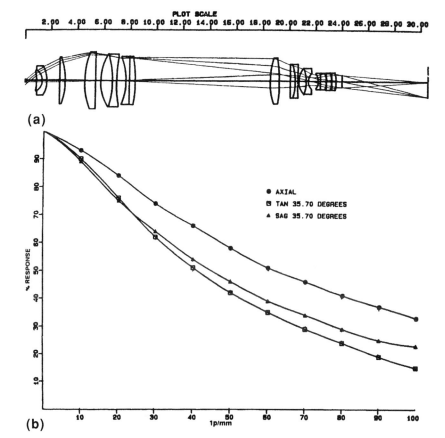

Figure 16-2 (a) 65-mm format periscope. (b) Lens MTF.

Table 16-3 Periscope Focusing (for 65 mm wide film)

Objd	T(14)	T(30)
infinity	10.148	6.576
18.92	10.448	6.276
9.82	10.748	5.976
6.78	11.048	5.675
5.26	11.348	5.375
4.35	11.648	5.075

Table 16-4 65-mm Format Periscope

	Radius	Thickness	Material	Diameter
0	0.00000	1.000E+07	Air	
1	−2.53781	0.4255	LAFN23	1.270
2	−0.78495	0.3603	SF1	1.400
3	−1.77334	0.9654	Air	1.980
4	0.00000	0.4220	LAKN12	3.160
5	−4.27376	1.4623	Air	3.160
6	4.82423	0.8338	LAKN12	3.860
7	0.00000	0.3837	Air	3.860
8	3.09811	0.8996	PK2	3.560
9	−5.31950	0.2935	SF1	3.560
10	7.95826	0.7202	Air	3.200
11	−3.77796	0.2617	SK4	3.200
12	0.00000	0.4529	SF1	3.260
13	−7.85088	10.1482	Air	3.260
14	4.47828	0.7039	SF5	3.110
15	−9.06419	0.8307	Air	3.110
16	0.00000	0.3743	FK51	2.280
17	−3.34290	0.1705	K7	2.280
18	12.13393	0.0933	Air	2.280
19	1.24162	0.5661	LAKN22	1.750
20	−2.08402	0.1537	SF5	1.750
21	0.70219	0.7595	Air	1.100
22	−0.76030	0.1545	SF5	1.000
23	12.32645	0.4786	LAKN22	1.130
24	−1.19560	0.0192	Air	1.130
25	−2.07104	0.3773	LAF2	1.080
26	−1.71724	0.0292	Air	1.180
27	−3.70720	0.3169	LAF20	1.080
28	−1.95831	0.4416	Air	1.180
29	Iris	6.5755	Air	0.824

The distance from the first lens surface to the image is 29.674. As with the previous periscope, the mirror is located at the entrance pupil, which is 0.900 in front of the first lens surface.

The lens element of FK51 glass helps to reduce secondary color. Distortion is 0.8%.

When computing the focusing movements for these periscopes (or any relay systems) it is important to check magnification changes as the lens is focused. That is, if during focus from long conjugate to a close distance, the relay magnification is reduced, the final image may not fill the film frame. One then has to start with a larger format to allow for this reduction in image size during focus.

Another type of cinematography periscope utilizes an available photographic front objective coupled with a field lens and unit power relay system (developed and manufactured by Century Precision, Burbank, CA). This has the advantage of excellent image quality, a relatively low $f^\#$, and most important the ability to use a wide variety of front objectives. The field lens is chosen to operate over a limited range of exit pupil locations. Although this excludes some lenses (fisheye) there are still many available that fall within the field lens correction. Its disadvantage is that the diameter of the photographic objective prevents the periscope from having its entrance pupil very close to the ground.

REFERENCES

1. Hajnal, S. (1980). Snorkel camera system, U.S. Patent #4195922.
2. Hopp, G. (1969). Periscop, U.S. Patent #3482897.
3. Kenworthy, P. (1973). A remote camera system for motion pictures, *SMPTE, 82:*159.
4. Kollmorgen, F. L. G. (1911). Periscope, U.S. Patent #1006230.
5. Laikin, M. (1980). Periscope lens systems. *American Cinematographer,* p. 702.
6. Latady, W. and Kenworthy, P. (1969). Motion picture camera system, U.S. Patent #3437748.
7. Samuelson, D. (1998). *Hands-On Manual for Cinematographers,* Focal Press, Oxford.
8. Taylor, W. (1977). Tri-power periscope head assembly, U.S. Patent #4017148.

17
IR Lenses

Infrared lenses differ from lenses designed for the visual region in several important aspects.

1. There are a lot fewer materials to choose from. Fortunately, available materials have high index of refraction and low dispersion (germanium, zinc selenide, etc.).
2. Due to the high cost of these materials and their relatively poor transmission, thickness should be kept to a minimum. Many of these materials are polycrystalline and exhibit some scattering: another reason to keep the lenses thin.
3. The long wavelength means a much lower resolution requirement.
4. The walls of the housing emit radiation and so contribute to the background.
5. Detectors are often linear arrays, in contrast to film or the eye. These detectors are usually cooled.
6. One must check that the detector is not being imaged back onto itself (narcissus, see Hudson, (1969), page 275; also see the discussion in Chap. 1).

Absorption in our atmosphere by H_2O, CO_2, and N_2O causes various windows or regions of transmission. Water vapor is the principal absorber in the 1 to 4 μm region, while carbon dioxide absorbs significantly at 2.7 μm and is also the main absorber between 4 and 5 μm. Therefore, the two main infrared windows are 3.2 to 4.2 μm and the

211

8 to 14 μm region (pp. 5–88 through 5–90 of the *Infrared Handbook*).

Infrared scanning systems are used for a wide variety of industrial and military purposes, some of which are cancer detection (mammograms), disclosing electronic problems in circuit boards, military night vision systems, and detection of hot bearings on railroad car wheels.

Other than the infrared region, single-element lenses are rarely used by themselves for imaging purposes. (The obvious exception is the meniscus lens used in inexpensive box cameras. Also see chap. 24 for single element laser focusing lenses.) However, in the IR, the high refractive index of germanium (and the generally lower resolution requirements of the IR) make the simple meniscus lens a useful device.

For a distant object and thin lens minimum spherical aberration (Riedl, 1974),

$$\frac{R_2}{R_1} = \frac{2N^2 + N}{2N^2 - 4 - N}$$

where R_1 and R_2 are the front and rear lens radii.

In Tables 17-1 and 17-2, values of R_1, R_2, T (the center thickness), and BFL are given for lenses of various materials. Effective focal length is 10, and the diameter is 3. Using the above formula, R_1 and R_2 were calculated and then a thickness chosen as to yield a 0.1 edge thickness. (The lens then had to be scaled to 10 EFL.)

Tables 17-1a and 17-1b were computed using the program BEST in Appendix E.

Figure 17-1 indicates how a FLIR (forward looking infrared) system operates. The objective forms an image onto a linear array of IR detectors. The amplified output of each detector drives a corresponding LED (light emitting diode). This array of LEDs is imaged, after reflection from the back side of the scanning mirror, onto an image tube (vidicon). This in turn drives a conventional cathode ray tube display. In this manner, a scanned infrared image is converted into a visual image onto a "TV" screen.

Table 17-1a Minimum Spherical Aberration Lenses at 3.63 μm[a]

R_1	R_2	T	BFL	Material
9.940	16.640	0.146	9.896	Silicon
8.934	23.994	0.180	9.882	As_2S_3
8.968	23.645	0.179	9.883	ZnSe
8.640	27.486	0.190	9.878	ZnS
10.213	15.249	0.137	9.899	Germanium

[a]Use the program Refractive Index in Appendix E to determine the refractive index.

Table 17-1b Minimum Spherical Aberration Lenses at 10.2 μm[a]

R_1	R_2	T	BFL	Material
10.203	15.297	0.137	9.899	Germanium
8.920	24.134	0.180	9.882	ZnSe
8.524	29.181	0.194	9.876	ZnS

[a]Use the program Refractive Index in Appendix E to determine the refractive index.

Figure 17-2 is a schematic of a typical FLIR objective. The front four lenses comprise an afocal telescope of magnification 9.41, which covers the spectral region 8 to 14 μm.

The distance from the first lens surface to the image is 16.314. The effective focal length is 25.057 (at 10.2 μm). In order to keep the diameter of the front lens to a minimum, the entrance pupil is placed in contact with this front element. The pupil diameter is 9.026 and so the f# is 2.776. This system was designed to use an array of 160 detectors, 0.004 inch on centers. Thus in the plane of the paper, the field of view is 1.47 deg.

The vertical-to-horizontal aspect ratio of typical TV display is 3/4 . The horizontal field of view of the objective then is 1.96 degrees. Since the effective focal length of the lenses to the right of the scan mirror is 2.66, the scan mirror moves a total of 9.276 degrees.

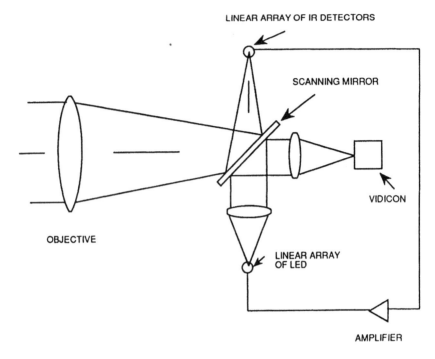

Figure 17-1 FLIR system.

The last two surfaces in the system represent a window used to protect the cooled detector array. Distortion at the top of the vertical field is 0.3%.

In some applications, the second through fourth lenses are mounted as a group. This group may then be moved out of the way and another group inserted to provide a wide field of view. The operator then may search the field in the wide FOV mode, and then, when the target is located, switch to narrow FOV.

Figure 17-3 shows an IR lens to be used in the 3.2 to 4.2 μm spectral region. It is f/1.5 and covers a 5 degree field. The EFL = 4.5 (at 3.63 μm).

Distortion is less than 0.1%. The distance from the first lens surface to the image is 6.834. An aperture stop should be located 0.135 be-

IR Lenses

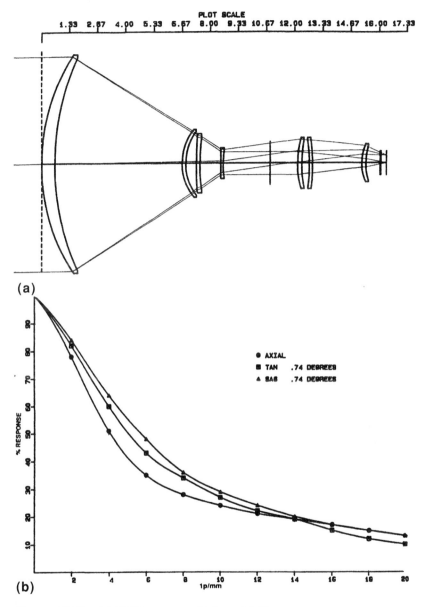

Figure 17-2 (a) FLIR systems. (b) Lens MTF.

Table 17-2 FLIR System

	Radius	Thickness	Material	Diameter
0	0.00000	1.00000E+08	Air	
1	7.68650	0.6227	Germanium	9.180
2	9.77390	5.9974	Air	9.000
3	2.06920	0.1753	Germanium	2.850
4	2.04300	0.4994	Air	2.700
5	12.77000	0.1750	ZnSe	2.500
6	6.93060	1.0077	Air	2.300
7	−46.96900	0.1000	Germanium	1.300
8	2.99040	2.2350	Air	1.200
9	0.00000	1.2909	Scan Mirror	1.800
10	3.39650	0.2000	Germanium	2.160
11	4.38450	0.3670	Air	2.050
12	−6.21400	0.1500	ZnSe	2.050
13	−7.13230	2.3190	Air	2.160
14	1.78200	0.1920	Germanium	1.600
15	2.14470	0.6745	Air	1.500
16	0.00000	0.0400	Germanium	1.000
17	0.00000	0.2682	Air	1.000

Table 17-3 IR Lens 3.2–4.2μm

	Radius	Thickness	Material	Diameter
0	0.00000	1.000E+07	Air	
1	8.56517	0.3249	Silicon	3.260
2	40.18391	1.0590	Air	3.160
3	−7.34648	0.2048	Germanium	2.300
4	−16.83771	2.0131	Air	2.380
5	3.87768	0.1571	Germanium	2.430
6	3.46173	0.1561	Air	2.300
7	18.52989	0.2179	Silicon	2.430
8	−11.30569	2.7012	Air	2.430

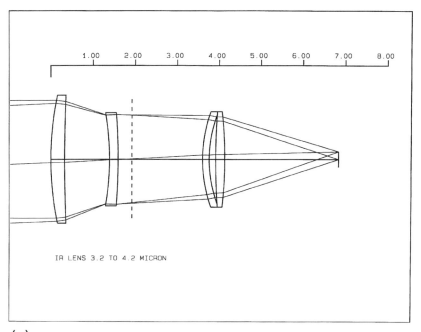

(a)

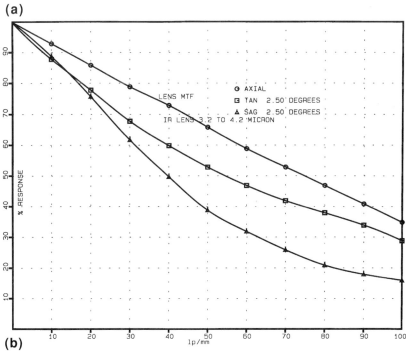

(b)

Figure 17-3 (a) IR 3.2–4.2 μm lens. (b) Lens MTF.

hind surface 4. The entrance pupil is located 1.9175 from the first lens surface. Detectors for the 3.2 to 4.2 micron region are often protected with a sapphire window. This was not included in the above design.

Sometimes it is desirable to expand a laser beam. This has the additional advantage that it also reduces the beam divergence by the magnification factor. (See Chap. 13 on afocal systems.) In Fig. 17-4 is shown such a beam expander for use with a CO_2 laser at 10.6 μm.

Table 17-4 10× Beam Expander

Surf	Radius	Thickness	Material	Diameter
0	0.0000	1.00000E+11	Air	
1	Stop	0.5000	Air	0.300
2	−0.2745	0.0500	ZnSe	0.300
3	0.0000	1.2921	Air	0.369
4	−2.4955	0.5000	ZnSe	2.824
5	−1.7706	0.0200	Air	3.010
6	−28.1662	0.3100	ZnSe	3.559
7	−6.1966		Air	3.594

In this case, the input beam intensity is assumed to have a Gaussian distribution of the form

$$I = I_0 e^{-2}$$

The intensity at the edge has fallen to 13.5% of the center. In the near IR region, 0.7 to 2.5 μm, ordinary glass can be used. Several very dense flints (Schott SF57 for example) that show considerable absorption in the visual region are very transparent here. In this spectral region, a very interesting thing occurs to the values of V (Walker, 1995). At 1.2 μm, V values of many crowns and flints are the same, while at 2.2 μm, V values of flints are often greater than those of the crowns.

Consider the region 2.0 to 2.5 μm, with center at 2.2 μm; then N and V values for BAK2 are 1.51566 and 72.355, while for SF4 we obtain 1.70972 and 97.084.

Schott (1992) has available a line of infrared transmitting glasses. These glasses exhibit good transmission from 0.4 μm to 4 μm (5 μm in the case of IRG11).

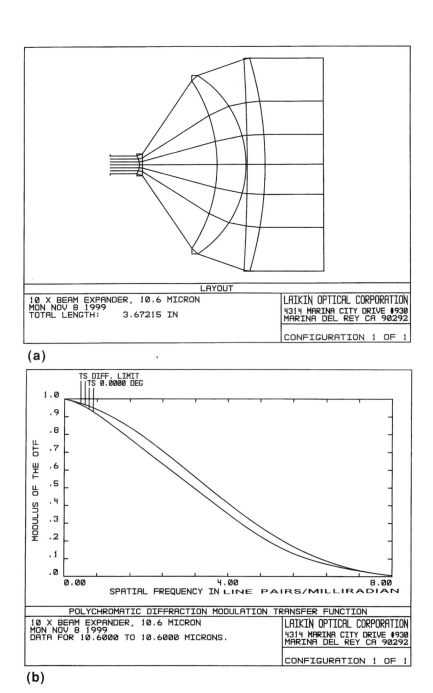

Figure 17-4 (a) 10× beam expander 10.6 μm. (b) Lens MTF.

219

Riedl (1996) presents a two-element (both of germanium) lens for the 8 to 12 μm region. It is a Petzval type of lens (the two elements have a large separation with the rear element being close to the image). To correct for the residual spherical aberration, the rear surface of the first element has a diffractive surface. This weak diffractive surface also improves the chromatic correction.

REFERENCES

Engineering Design Handbook, Infrared Military Systems (1971). AD 885227, available from NTIS, Springfield, VA.

Optical Materials for Infrared Instrumentation, IRIA (1959). University of Michigan report #2389, also the supplement of 1961, Information and Analysis Center, Ann Arbor, MI.

Fjeldsted, T. P. (1983). Four element infrared objective lens U.S. Patent #4380363.

Hudson, R. (1969). *Infrared System Engineering,* John Wiley, New York.

Jamieson, T. H. (1976). Aplanatic anastigmatic two and three element objectives, *Applied Optics, 15:*2276.

Kirkpatrick, A. D. (1969). Far infrared lens, U.S. Patent #3439969.

Kokorina, V. F. (1996). *Glasses for Infrared Optics,* CRC Press, Boca Raton, FL.

Lloyd, J. M. (1975). *Thermal Imaging Systems,* Plenum Press, New York.

Neil, I. A. (1985). Infrared objective lens system, U.S. Patent #4505535.

Norrie, D. G. (1986). Catadioptric afocal telescopes for scanning infrared systems, *Optical Engineering, 25:*319.

Riedl, M. J. (1974). The single thin lens as an Objective for IR imaging systems, Electro-Optical Sys. Design, p. 58, M.S. Kiver.

Riedl, M. J. (1996). Design example for the use of hybrid optical elements in the IR, OPN Engineering and Laboratory Notes.

Rogers, P. J. (1977). Infrared lenses, U.S. Patent #4030805.

Schott (1992). Data sheet #3112e, Schott Glass Technologies, Durea, PA.

Sijgers, H. J. (1967). Four element infrared objective, U.S. Patent #3321264.

Wolf, W. and Zissis, G. (1985). *The Infrared Handbook,* Research Institute of Michigan, Ann Arbor, MI.

18
Ultraviolet Lenses and Optical Lithography

Lenses for the UV spectral region are used in photo-mask printing, printing of reticles onto a material with a photo-resist, microscope objectives for fluorescence studies (see Fig. 11-4), forensic photography, optical lithography, etc.

As with the infrared spectral regions, a major problem is finding suitable materials. For the spectral region 0.2 to 0.4 µm, the most important materials are fused silica and calcium fluoride. Although several glasses have been developed to have reduced absorption below 0.4 µm, they have limited use since there is considerable absorption even at 0.3 µm. One such glass is UBK7. Refractive index data for this and other materials can be computed using the Refractive Index program in Appendix E. At 0.32 µm wavelength and a sample thickness of 25 mm, UBK7 has an internal transmittance (ignoring surface reflection) of 0.66.

Recently, Schott (ULTRAN 30, Schott, 1992) and Ohara (i-line glasses, Ohara, 1993) have announced a new family of UV transmitting glasses. These materials have high chemical durability as well as being transparent over a large wavelength region. Table 18-1 lists these new materials with their refractive index, dispersion, and wavelength at which a 25 mm thickness will transmit 70%.

Schott has developed a series of i-line glasses. These glasses have the same refractive index values as the regular glasses but are melted and tested to have high transmittance at 0.365 µm. These

221

Table 18-1 New UV Transmitting Glasses

Glass	N_d	V_d	Wavelength μm for 70% transmittance
ULTRAN 30	1.5483	74.2	0.282
S-FPL51Y	1.4970	81.1	0.310
S-FSL5Y	1.4875	70.2	0.302
BSL7Y	1.5163	64.1	0.315
BAL15Y	1.5567	58.7	0.323
BAL35Y	1.5891	61.2	0.318
BASM51Y	1.6031	60.7	0.320
PBL1Y	1.5481	45.9	0.319
PBL6Y	1.5317	49.0	0.316
PBL25Y	1.5184	40.8	0.336
PBL26Y	1.5673	42.9	0.337
PBM2Y	1.6200	36.3	0.345
PBM8Y	1.5955	39.3	0.339

glasses are FK5HT, BK7HT, K5HT, K7HT, LLF1HT, LLF6HT, LF5HT, LF6HT, F2HT, F8HT, and F14HT.

Sapphire is often used as a window material for certain types of detectors. Antireflection coatings are a problem because some commonly used coating materials are absorbing in this spectral region. The usual reflection curves, which the coater presents, are not adequate here. Always insist on a transmission curve as well.

Figure 18-1 shows a lens designed for this spectral region (0.2 μm to 0.4 μm with center at 0.27 μm). It has a focal length of 5, is f/4, and has a FOV of 10 degrees.

Figure 18-2 shows a Cassegrain lens system used to image an object 2.75 in diameter at a magnification of 0.1 (0.2 μm to 0.4 μm with center at 0.27 μm).

Note that the secondary mirror (R= 1.5997) has a different radius from the rear corrector surface. This means that the corrector/secondary has to be made in two sections, the secondary being aligned and bonded to the corrector. The primary mirror is a Mangin type. In the visual region, silver (suitably protected with an overcoat of copper

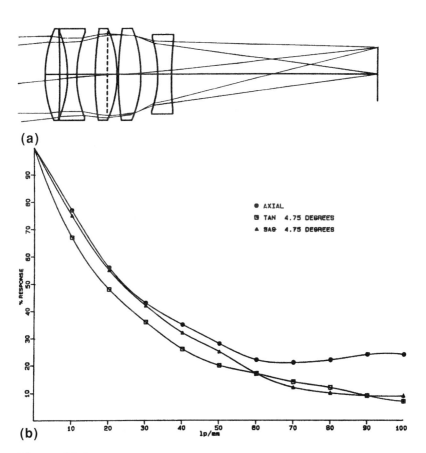

Figure 18-1 (a) Quartz and CaF$_2$ ultraviolet lens. (b) Lens MTF.

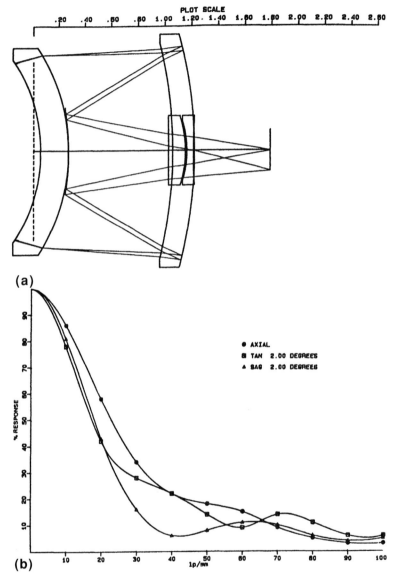

Figure 18-2 (a) Cassegrain objective, all-quartz. (b) Lens MTF.

Table 18-2 Quartz and CaF_2 UV Lens

	Radius	Thickness	Material	Diameter
0	0.00000	1.00E+07	Air	
1	1.81630	0.2688	CaF_2	1.500
2	−9.70194	0.1363	Air	1.500
3	−1.71562	0.1619	Quartz	1.400
4	1.72652	0.3324	Air	1.400
5	4.26856	0.3909	CaF_2	1.500
6	−1.73853	0.0223	Air	1.500
7	4.39230	0.4152	CaF_2	1.500
8	−1.78979	0.3056	Air	1.500
9	−1.70410	0.2496	Quartz	1.200
10	5.60739	3.7241	Air	1.200

The distance from the first lens surface to the image is 6.007. Distortion is 0.2%. The entrance pupil is located 1.114 from the first lens surface.

Table 18-3 Cassegrain Objective

	Radius	Thickness	Material	Diameter
0	0.00000	39.370	Air	
1	−0.98911	0.2134	Quartz	1.230
2	−1.14451	0.7869	Air	1.390
3	−2.75430	0.1617	Quartz	1.460
4	−2.74164	−0.1617	Mirror/Quartz	1.570
5	−2.75430	−0.7869	Air	1.460
6	−1.59975	0.7566	Mirror	0.580
7	−13.92287	0.1242	Quartz	0.470
8	−0.69773	0.0110	Air	0.470
9	−0.64031	0.0480	Quartz	0.410
10	3.05088	0.5769	Air	0.470

The distance from the first lens surface to the image is 1.730. Distortion = 0.7%. $f^\# = 3$. (NA = 0.1515.) Entrance pupil of 1.191 diameter is 0.056 in front of the first lens surface.

and then painted) is generally used as a second surface mirror because of its higher reflectivity than aluminum. However, in the 0.2 to 0.4 µm region, aluminum is the preferred coating material, since silver has reduced reflectivity below 0.38 µm (Hass, 1965).

Optical lithography is used in the production of computer chips (Mack, 1996). This typically involves a very high resolution lens operating at a single UV wavelength. For a circular uniformly illuminated pupil the radius of the Airy disc is

$$R = \frac{0.61\lambda}{NA}$$

Inside this radius is 84% of the energy. Thus to increase the resolving power we need to reduce the wavelength and use a high NA lens system. In order to obtain this high resolution, it is helpful to have a single wavelength source. Many years ago, this was accomplished by the use of one of the spectral lines (0.365 µm, 0.405 µm, and 0.436 µm) of a mercury arc lamp. To obtain higher resolution it is necessary to use even shorter wavelengths. Today, the preferred source is an eximer laser. Eximer is short for excited dimer, a molecule consisting of two identical atoms such as He_2 or Xe_2 that exist only in the excited state.

This term now includes ArF, KrCl, KrF, XeCl, and XeF.

Laser	Wavelength (µ)
F_2	0.157
ArF	0.193
KrCl	0.222
KrF	0.248
XeCl	0.308
XeF	0.350

Unfortunately, there are only a few materials that are transparent at wavelengths below 0.35 µm. These are calcium fluoride, lithium fluoride, magnesium fluoride, sapphire, and fused silica. Transmission at these short wavelengths is dependent on material purity. Dam-

Ultraviolet Lenses and Optical Lithography

age threshold, index homogeneity, and fluorescence are additional problems.

Polishing of optics for high-energy UV optics is also a problem. A λ/10 surface specification at 0.193 is nearly three times tighter than the same specification at 0.546 μm. Subsurface damage has also been identified. During the act of polishing the surface, there is a layer of material that "flows" while the material is being worked. This causes microfractures and other surface defects that under high illumination levels eventually cause failure.

The manufacture of integrated circuits on a silicon substrate involves a step-and-repeat process. This requires an optical system with very low distortion that generally must be telecentric in both object and image space. Such a system is shown in Fig. 18-3. It was designed for use with a KrF 0.248 μm source. The object is 93.6 mm in diameter and the magnification is –0.25. NA is 0.56. This is a minor optimization from the Sasaya (1998) patent.

Table 18-4 Lithography Projection Lens

Surf	Radius	Thickness	Material	Diameter
0	0.0000	4.2403		3.685
1	–22.1726	1.1965	Silica	4.724
2	–8.0708	0.0340		4.948
3	7.8597	2.8011	Silica	4.979
4	–16.7458	0.0197		4.536
5	17.0155	0.2780	Silica	4.393
6	6.9624	0.5206		4.215
7	–15.0217	0.2737	Silica	4.203
8	9.9267	1.2230		4.154
9	–118.0014	0.4544	Silica	4.254
10	8.9161	1.3041		4.314
11	–3.2403	0.4537	Silica	4.386
12	22.5155	0.3582		5.602
13	218.4297	1.4444	Silica	5.952
14	–6.1734	0.0357		6.441
15	24.1039	1.6301	Silica	7.441
16	–8.7303	0.0296		7.623
17	20.9338	1.2186	Silica	7.765

Table 18-4 Continued

Surf	Radius	Thickness	Material	Diameter
18	−22.3623	0.0219		7.737
19	16.6893	0.8660	Silica	7.529
20	−55.2167	0.0229		7.426
21	8.0228	0.8944	Silica	6.883
22	32.8470	2.6744		6.680
23	−29.0004	0.3302	Silica	4.250
24	4.1066	0.9537		3.766
25	−7.1881	0.4344	Silica	3.679
26	11.4624	2.2895		3.649
27	−2.9371	0.4466	Silica	3.806
28	Stop	0.4483		4.512
29	−11.1591	0.8745	Silica	4.779
30	−5.6000	0.0546		5.314
31	−228.0996	1.5604	Silica	6.157
32	−5.7742	0.0357		6.567
33	25.7835	1.4195	Silica	7.149
34	−13.6893	0.0210		7.279
35	10.0326	1.2422	Silica	7.291
36	84.2350	0.0328		7.132
37	6.4925	1.0965	Silica	6.819
38	13.7781	0.0317		6.497
39	4.2927	2.8749	Silica	5.923
40	2.9789	2.1318		3.541
41	1.8260	0.6676	Silica	1.973
42	3.9400	0.5148		1.525
43	0.0000	0.0000		0.921

The effective focal length is 78.16. The object-to-image distance is 39.454.

Note: The diameters listed are clear apertures as required by ray tracing, not actual lens diameters as indicated in the other lens prescriptions. Distortion and field curvature plots are shown in Fig. 18-3c. Maximum distortion occurs at about 0.75 of full field. The separation between the sagittal and tangential foci indicates that there is some small astigmatism present.

In order to obtain even smaller image sizes, the trend is toward using source wavelengths shorter than 0.193 μm (Kubiak and Kania

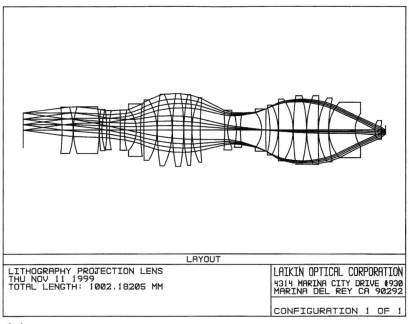

(a)

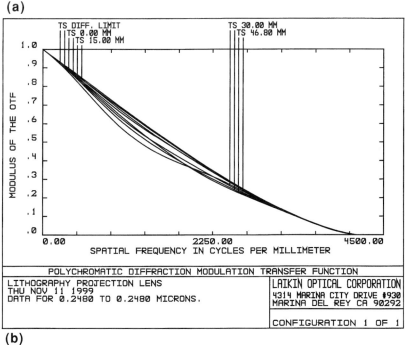

(b)

Figure 18-3 (a) Lithography projection lens. (b) Lens MTF. (c) Distortion and field curvature plots.

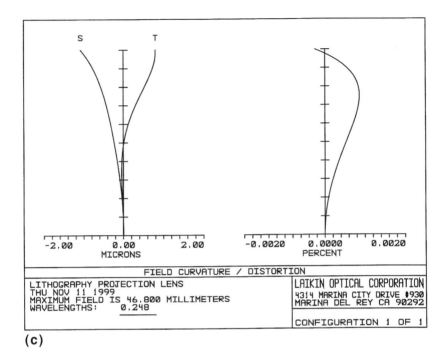

(c)

Figure 18-3 Continued.

1996). All reflective optics in a vacuum or helium environment are then required.

REFERENCES

Betensky, E. I. (1967). Modified Gauss type of optical objective, U.S. Patent #3348896.

Elliott, D. J. and Shafer, D. (1996). Deep ultraviolet microlithography system, U.S. Patent #5488229.

Hass, G. (1965). Mirror Coatings, Chapter 8 of *Applied Optics and Optical Engineering* (Kingslake, R., ed.) *3:*309, Academic Press, New York.

Hwang, S. (1994). Optimizing distortion for a large field submicron lens, *SPIE, 2197:*882.

Kubiak, G. D. and Kania, D. R., eds. (1996). *OSA trends in Optics and Photonics,* Vol. 4, *Extreme Ultraviolet Lithography,* Optical Society of America, Washington, D.C.

Lowenthal, H. (1970). Lens system for use in the near ultraviolet, U.S. Patent #3517979.

Mack, C. A. (1996). Trends in optical lithography, *Optica and Photonics News,* April, 29.

Ohara (1993). *i-Line Glasses,* Ohara Corp., Somerville, New Jersey.

Sasaya, T. Ushida, K., and Mercado, R. (1998). All fused silica 248 nm lithographic projection lens, U.S. Patent #5805344.

Sheats, J. R. and Smith, B. W., eds. (1998). *Microlithography: Science and Technology,* Marcel Dekker, New York.

Schott (1992). Schott Glass Catalog, Schott Glass Technologies, Durea, Penn.

Tibbetts, R. E. (1962). Ultra-violet lens, U.S. Patent #3035490.

19
F Theta Scan Lenses

These lenses are used to scan a document for either printing or reading. At an external entrance pupil is a scanning device, perhaps a rotating polygon, rotating mirror, mirror galvanometer, piezoelectric deflector, etc. The document to be scanned is usually flat. The rotating polygon rotates at a high uniform angular velocity. Therefore for the image positions to be uniformly spaced it is necessary that the image height be proportional to the scan angle (not to the tangent of the angle as in a photographic lens).

Image height = $KF\theta$

where F is lens focal length and θ is the scan angle. If θ is in degrees, then $K \approx 0.0175$.

In figure 19-1 is shown a document scan lens, 24 inch focal length, 2 inch diameter entrance pupil, visual correction, designed to scan a 14 inch wide document for a scan angle of 33.4 degrees.

The entrance pupil is located 1.5015 in front of the first lens surface. The distance from the first lens surface to the image is 29.722. As one would expect, resolution at the edge of the field is limited by lateral secondary color. Surface 7 could have been made a plane surface. The edge thickness on the last lens is a little thin. Following is a list of image heights vs. scan angle.

θ degrees	Image height
5.561	2.3301
11.122	4.6715
16.7	7.0398

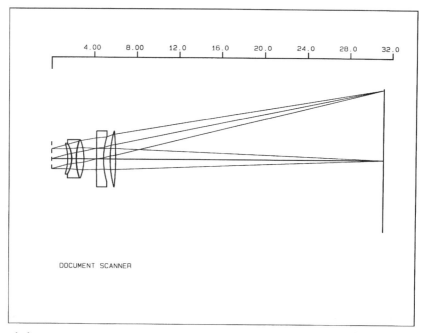

(a)

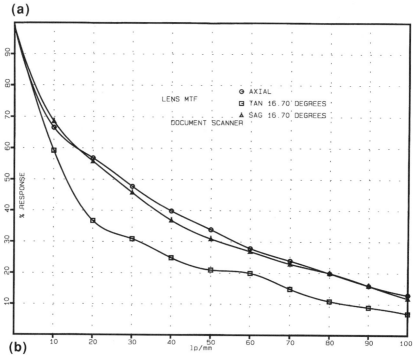

(b)

Figure 19-1 (a) Document Scanner. (b) Lens MTF.

Table 19-1 Document Scanner

	Radius	Thickness	Material	Diameter
0	0.00000	1.000E+08	Air	
1	−5.04096	0.3554	LAKN22	2.900
2	−3.42954	0.0315	Air	3.080
3	−3.04156	0.4005	ZKN7	3.040
4	6.22287	0.0317	Air	3.840
5	6.55330	0.7052	LAKN22	3.840
6	−5.58009	1.1996	Air	3.840
7	363.61124	0.6667	SF1	5.610
8	8.48491	0.6158	Air	4.540
9	11.21622	0.5155	LAKN22	5.610
10	−28.91675	25.2000	Air	5.610

In figure 19-2 is shown a scan lens for use with an argon laser (0.488 micron). The focal length is 20 inches and it is f/30. The scan angle is 30 degrees.

Table 19-2 Argon Laser Scan Lens

	Radius	Thickness	Material	Diameter
0	0.00000	1.000E+08	Air	
1	−1.18815	0.1221	F4	1.170
2	−1.24034	0.0232	Air	1.300
3	−3.95935	0.1409	SF18	1.300
4	−2.96846	0.5432	Air	1.440
5	−1.33106	0.1617	F5	1.520
6	−1.43429	21.1878	Air	1.700

The entrance pupil is 1.000 in front of the first lens surface and is 0.667 in diameter. The distance from the first lens surface to the image is 22.179.

In figure 19-3 is shown a scan lens for use at the 0.6328 μm (HeNe) laser line. The effective focal length is 2 and is f/3. It is a modification of the system presented in Murthy (1992).

The distance from the entrance pupil to the image surface is

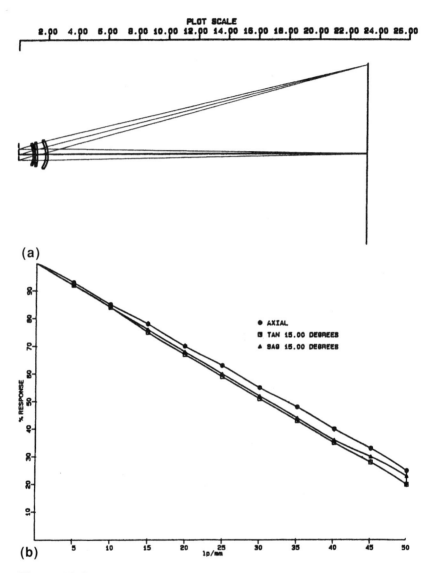

Figure 19-2 (a) Argon laser scan lens. (b) Lens MTF.

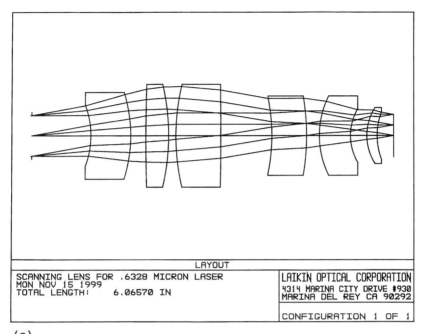

(a)

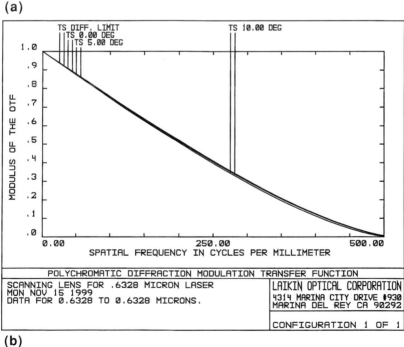

(b)

Figure 19-3 (a) Scanning lens for 0.6328 μm laser. (b) Lens MTF.

Table 19-3 Scanning Lens for 0.6238 μm Laser

Surf	Radius	Thickness	Material	Diameter
0	0.0000	1.000E+11		
1	Stop	1.0000		0.667
2	−1.3472	0.6742	SF57	1.000
3	−2.2750	0.2541		1.420
4	31.2679	0.3753	SF57	1.680
5	−3.6913	0.1199		1.680
6	3.6146	0.7496	SF57	1.680
7	−47.9697	0.8578		1.680
8	−2.9183	0.5843	SF57	1.240
9	−3.7040	0.2148		1.300
10	1.4362	0.5219	SF57	1.300
11	0.8715	0.2686		0.880
12	0.8596	0.1681	SF57	0.920
13	1.0989	0.2773		0.780
14	0.0000	0.0000		0.697

6.066 As can be seen from the MTF plot in Fig. 19-3b, the lens is diffraction limited over the full field.

Following is a list of image heights vs. scan angle.

θ degrees	Image height
2	0.0698
4	0.1394
6	0.2091
8	0.2786
10	0.3483

REFERENCES

Beiser, L. (1974). Laser Scanning Systems, Chapter 2, Vol. 2, *Laser Applications* (M. Ross, ed.), Academic Press, New York.

Brixner, B. and Klein, M. M. (1992). Optimization to create a four-element laser scan lens from a five element design, *Optical Engineering,* 31:1257.

Buzawa, M. J. and Hopkins, R. E. (1975). *Optics for laser scanning, SPIE, 53*:9.

Cobb, J. M., LaPlante, M. J., Long, D. C., and Topolovec, F. (1995). Telecentric and achromatic F-theta scan lens, U.S. Patent #5404247.

Fisli, T. (1981). High efficiency symmetrical scanning optics, U.S. Patent #4274703.

Griffith, J. D. and Wooley, B. (1997). High resolution 2-D scan lens, U.S. Patent #5633736.

Hopkins, R. (1987). Optical system requirements for laser scanning systems, *Optics News,* p. 11.

Maeda, H. (1983). An F-θ lens system, U.S. Patent #4401362.

Maeda, H. (1984). An F-θ lens system, U.S. Patent #4436383.

Marshall, G. (1985). *Laser Beam Scanning,* Marcel Dekker, New York.

Minami, S. (1987). *Scanning optical system of the Canon laser printer, SPIE, 741*:118.

Murthy, E. K. (1992). Elimination of the thick meniscus element in high-resolution scan lenses, *Optical Engineering, 31:*95.

Starkweather, G. (1980). *High Speed Laser Printing Systems,* Laser Applications 4:125 (M. Ross, ed.), Academic Press, New York.

20
Endoscopes

Endoscopes are very long and narrow optical relay systems designed to probe the interior parts of the body. The object medium then is water. Therefore the front element is usually flat, permitting the endoscope to view objects in air or in the body. Because of their long and narrow arrangement, endoscopes are also used in industry as a means of inspecting the interior parts of deep bores. They are then called borescopes.

In some systems, a front objective is used to image the specimen onto a fiber-optic bundle. This flexible bundle then transfers the image to its opposite end where it is viewed with an eyepiece. A separate fiber-optic bundle is also used to provide illumination. This provides a flexible system and is often preferred for medical examinations.

Rigid systems transfer the image by means of a series of relay/field lenses. Image quality is generally superior to the fiber-optic system (Tomkinson, 1996). A fiber-optic bundle is used for illumination purposes.

Some of the earlier endoscopes were designed by separately designing the front objective and a relay. Several identical relay units were then assembled with the front objective to form a complete system. The problem with such a technique is that each relay is not completely corrected. When four or more such units are assembled, the aberrations add. All modern lens design programs have the ability to tie surfaces together. Thus the entire system can be designed as a complete system. For example, in the following design, the curvature of surfaces 9, 15, and 21 were tied together. In this mode, the curvature of surface 9 is an independent variable, and then curvatures 15 and 21 are set equal to it; likewise for surfaces 10, 16, and 22.

In figure 20-1 is shown an endoscope using rod lenses. Such lens elements are very long in relation to their diameter. This helps in the mounting and assembly of the instrument. The thick lens feature also helps to reduce the Petzval sum as well as spherical aberration. It has a field of view in water of 50 degrees and is f/3 (see Hopkins, 1966).

The distance from the first lens surface to the image is 8.153. The effective focal length is 0.0669. Distortion = 4.6%. Note that with the exception of the last surface, the radii, thicknesses, and materials of the relay lenses are repeated.

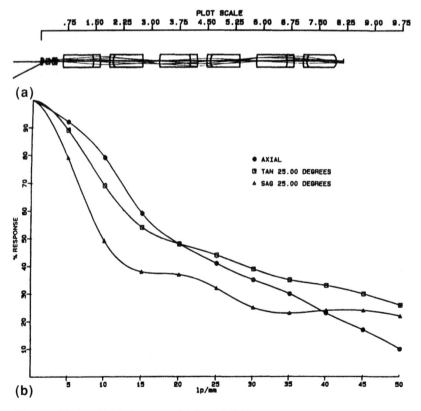

Figure 20-1 (a) Endoscope. (b) Lens MTF.

Table 20-1 Endoscope

	Radius	Thickness	Material	Diameter
0	0.00000	1.000E+07	Water	
1	0.00000	0.0310	BK7	0.160
2	0.06157	0.0999	Air	0.090
3	Stop	0.0076	Air	0.041
4	0.28205	0.1139	LAKN12	0.130
5	−0.18104	0.0574	Air	0.130
6	0.16108	0.1050	SK16	0.160
7	−0.06792	0.0321	SF4	0.134
8	−0.32075	0.1635	Air	0.160
9	0.58699	0.8304	LAFN21	0.300
10	−0.32643	0.1884	SF4	0.300
11	−0.69376	0.2429	Air	0.300
12	0.64588	0.0969	SF4	0.300
13	0.21519	0.7971	LAFN21	0.300
14	−1.69480	0.4423	Air	0.300
15	0.58699	0.8304	LAFN21	0.300
16	−0.32643	0.1884	SF4	0.300
17	−0.69376	0.2429	Air	0.300
18	0.64588	0.0969	SF4	0.300
19	0.21519	0.7971	LAFN21	0.300
20	−1.69480	0.4423	Air	0.300
21	0.58699	0.8304	LAFN21	0.300
22	−0.32643	0.1884	SF4	0.300
23	−0.69376	0.2429	Air	0.300
24	0.64588	0.0969	SF4	0.300
25	0.21519	0.7971	LAFN21	0.300
26	−0.19603	0.1914	Air	0.300

REFERENCES

Hoogland, J. (1986). Flat field lenses, U.S. Patent #4575195.
Hopkins, H. H. (1966). Optical system having cylindrical rod-like lenses, U.S. Patent #3257902.
Nakahashi, K. (1981). Optical system for endoscopes, U.S. Patent #4300812.
Nakahashi, K. (1983). Objective for endoscope, U.S. Patent #4403837.
Ono, K. (1998). Image transmitting optical system, U.S. Patent #5731916.

Rol, P., Jenny, R., Beck, D., Frankhauser, F., and Niederer, P. F. (1995). Optical properties of miniaturized endoscopes for opthalmic use, *Optical Engineering, 34:*2070.

Tomkinson, T. H., Bentley, J. L., Crawford, M. K., Harkrider, C. J., Moore, D. T., and Rouke, J. L. (1996). Ridgid endoscopic relay systems: a comparative study, *Applied Optics, 35:*6674.

Yamashita, N. (1977). Retrofocus type objective for endoscopes, U.S. Patent #4042295.

Yamashita, N. (1978). Optical system for endoscope, U.S. Patent #4111529.

21
Enlarging Lenses

These lenses all show some degree of symmetry and are used to project film onto sensitized paper. They generally cover magnification ratios 2–8 or 8–16. The later case is for enlarging 35 mm SLR film to 8 × 10 or 11 × 14 inch prints. f#s are usually f/2.8 or larger.

Due to the sensitivity of many photographic papers to wavelengths shorter than 0.4 µm, it is desirable to select glasses such that any energy below this wavelength is absorbed. (It is also very difficult to correct lens aberrations at these shorter wavelengths.) Conversely, the system must show high transmission at wavelengths longer than 0.4 µm.

Figure 21-1 shows an enlarging lens designed for 10× magnification. It is designed to cover a 44 mm diagonal image (35 mm SLR format): The focal length is 2.566. Distortion at the edge is 0.4%.

This double Gauss type of design would represent a premium lens for the professional photographer. A lower priced lens would be a Tessar type of design (of the type shown in Fig. 4-2) or, for even lower cost, a three-element f/5.6. This is the most popular type for amateur use. It yields its best performance when stopped down to f/8. Most enlarging lenses have an iris with a maximum f# of 32. Focal lengths longer than 150 mm generally have an iris to cover the range f/5.6 to 45.

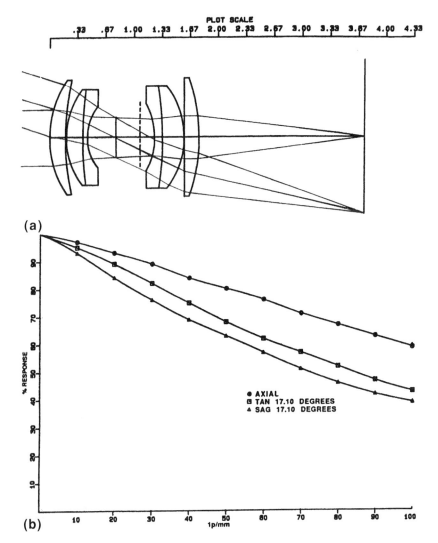

Figure 21-1 (a) 65-mm, f/4 10× enlarging lens. (b) Lens MTF.

Enlarging Lenses

Table 21-1 10× Enlarging Lens

	Radius	Thickness	Material	Diameter
0	0.00000	26.5300	Air	
1	0.99475	0.1893	LAF2	1.220
2	2.56734	0.0116	Air	1.220
3	0.78085	0.1966	PK2	1.020
4	3.66775	0.0553	F4	1.020
5	0.49146	0.3420	Air	0.660
6	Iris	0.4509	Air	0.420
7	−0.67335	0.0963	SF1	0.720
8	−2.86335	0.2556	LAKN7	1.090
9	−0.81751	0.0058	Air	1.090
10	−32.38740	0.1707	LAF2	1.260
11	−1.70995	1.9624	Air	1.260

The distance from the first lens surface to the image is 3.736.
The entrance pupil distance from the first lens surface is 1.072.

REFERENCES

Kouthoofd, B. J. (1998). Enlarging lens with a large working distance, U.S. Patent #5724191.

Matsubara, M. (1977). Enlarging lens system, U.S. Patent #41045127.

Matsubara, M. (1977). Enlarging lens system, U.S. Patent #4057328.

Matsubara, M. (1977). Enlarging lens system, U.S. Patent #4013346.

Velesik, S. (1975). Reproduction lens system, U.S. Patent #3876292.

22
Projection Lenses

Although it is true that a camera lens can be used as a projection lens, projection lenses do have several differences from camera lenses.

1. They must be able to withstand the high power densities encountered in projection. (This assumes film projection. See the discussion for Fig. 22-6, an LCD projection lens.) This is particularly true when a high-wattage xenon arc is used. There is usually no problem with coatings; the problem is with optical cements. Canada balsam is totally unsuitable. Fortunately, this is rarely used today. Lens Bond, a thermosetting optical cement made by Summers Labs (Fort Washington, PA), works very well. It is a polyester resin. Versions are available for room temperature curing, oven cure, and cure with a UV lamp. They meet the requirements of MIL-A-3920. $N_d = 1.55$. A similar material is made by Norland Optical (New Brunswick, NJ).

Epoxy cements are also sometimes used. These materials can withstand very high temperatures. A material that the author has found to be useful is Hysol OSO-100. Parts must be cured overnight at 100 deg C. $N_d = 1.493$. Lenses can withstand temperatures to 125 deg C. Another useful material is TRA-BOND F114, Tracon Inc., Medford, MA, $N_d = 1.54$. This is useful over the temperature extremes –60°F to 130°F after a 24 hour room temperature cure.

2. An iris is not needed. These lenses have a fixed aperture stop.

3. Sufficient back focus must be provided to clear the film transport mechanism. This generally eliminates a field-flattening lens near the film plane (ANSI, 1982).

4. The lens exit pupil must correspond to a source image of the lamp. This is typically 4 inches from the film plane for xenon arc projectors (70 mm projection; somewhat less for Academy (35 mm) format projection).

5. Since the projection distance is always at least 100 focal lengths, projection lenses are frequently designed for infinite conjugates. This can sometimes be a dangerous mistake, since for wide-angle lenses, with appreciable distortion, the lens aberrations are strongly influenced by conjugate distances. See the comments regarding this in Chap. 9. All designs in this section are presented for an infinite conjugate, and the user is advised to check performance at the actual conjugate distance.

6. The screen is generally a cylinder and curved toward the audience. This radius is usually between 0.8 to 1.5 times the lens-to-screen distance. On a first order basis, the film should be on a radius the same as the screen radius. However, since this radius is very long as compared to the film width, a flat film surface is generally used. Projectors made by Pioneer have a 22 inch cylindrical radius (the center of curvature is toward the arc). Strong-Ballantine projectors have a flat gate.

7. ANSI 196M calls for 16 ft lamberts screen illuminance. Although the eye resolves 1 minute of arc at 100 ft lamberts, its resolution drops to 2 minutes of arc at 16 ft lamberts screen illuminance. If an observer were in the rear of the theater (at the projector), his angular resolution would be

$$\theta = \frac{1}{RF}$$

where R is the resolution at the film in lp/mm and F the focal length of the projection lens (in mm). This applies to an observer at the rear. At the front of the theater an observer will always see an aberrated image (if the lens were perfect, he would see the magnified image of the film grain). A compromise then is to achieve a resolution at the middle of the theater of 2 minutes of arc. This is equivalent to

Projection Lenses

$$\text{one minute of arc} = \frac{1}{RF}$$

at the center of the field of the projection lens. At 30° off axis (as seen by an observer at the center of the theater), the resolution could be reduced to 5 minutes of arc.

The Projection Lens program in Appendix E is helpful in determining the lens focal length and lens shift four various projection distances and format. Another very useful program, Theater Design Pro 2. lb from Schneider Optics, is available as a download on the Internet.

Figure 22-1 shows a 3 inch FL f/1.8 lens designed to project 35 mm motion picture film (1.07 diagonal).

Table 22-1 3.0 Inch $f/1.8$ Projection Lens

	Radius	Thickness	Material	Diameter
0	0.00000	1.000E+07	Air	
1	2.29280	0.4300	SKI6	2.600
2	7.62805	0.0155	Air	2.500
3	1.19781	0.5376	SK2	2.060
4	2.86446	0.1346	F4	1.760
5	0.71271	0.6649	Air	1.240
6	Stop	0.2231	Air	0.950
7	−1.02332	0.1262	F4	1.010
8	1.38177	0.6049	SK2	1.540
9	−1.28556	0.0127	Air	1.540
10	2.70103	0.3803	SK16	1.700
11	−3.63592	1.8022	Air	1.700

The distance from the first lens surface to the film is 4.932. Distortion is 1.7%. The entrance pupil is 2.786 from the first lens surface.

In most movie theaters, the projector is nearly at the screen center line. However, there are times when the projector is substantially above the screen center line. For certain situations, having the projector above the audience allows a more efficient theater design. Unfortunately, it

252 Chapter 22

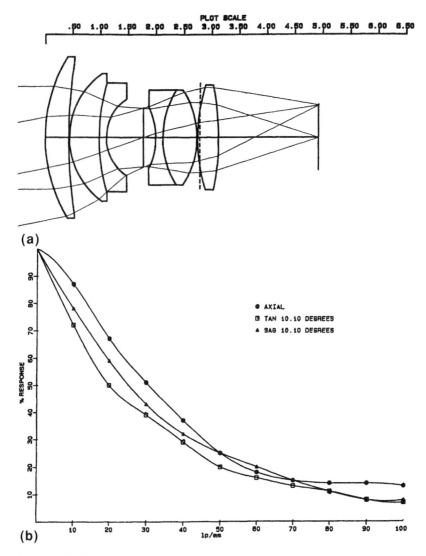

Figure 22-1 (a) 3.0-in., *f*/1.8 projection lens. (b) Lens MTF.

Projection Lenses 253

causes keystone distortion: a square is projected as a trapezoid. For some cases this may not be objectionable. It may be eliminated by displacing the lens with respect to the film center line. This is shown in Fig. 22-2.

You will note that the lens is "dropped" a distance D below the film centerline.

$$D = \frac{FH}{S-F}$$

Although this arrangement has the desirable effect of eliminating keystone distortion, it means that the projection lens must be designed for a larger field of view than normal. The Kodak 3D theater at the EPCOT center in Florida uses this principle. In this case two projectors 5.5 feet (horizontal) apart are used to project 70 mm film. Each projector has a Polaroid sheet in front of the lens, and the viewers wear polarizing glasses. Both images must be accurately registered over the entire screen. However, due to the projector sepa-

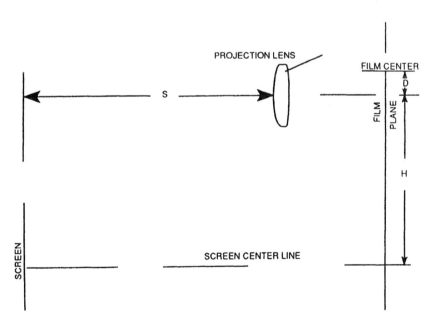

Figure 22-2 Displaced projection lens.

ration, distortion at the edge of the field will be different for each projector.

$S = 1236$
$F = 3.594$
$H = 33$
$D = 0.096$

Referring to Appendix A, we see that the standard for 65 mm film is $1.913 \times .868$. Due to the lens displacement (in this case a horizontal displacement), the lens has to cover an additional horizontal amount of $2D$. This corresponds to a diagonal of 2.277.

This f/2 lens is shown in Fig. 22-3.

Table 22-3 Projection Lens for 70mm Film

Surf	Radius	Thickness	Material	Diameter
0	0.0000	1.00E+10		
1	5.4309	0.7220	FK5	4.460
2	2.3942	1.2934		3.620
3	−3.8629	0.7140	SF5	3.620
4	−5.5409	0.0115		4.460
5	−16.0642	1.6219	LAK9	3.940
6	−6.5538	0.0134		4.460
7	4.3769	1.6521	LAK10	4.460
8	−4.1619	0.4850	F2	4.460
9	9.8038	1.5399		3.000
10	Stop	0.5755		1.732
11	−3.8829	0.1769	SF4	1.920
12	5.4280	0.0844		2.080
13	35.4552	0.3503	LAK10	2.280
14	−3.2140	0.0100		2.280
15	3.2160	0.6218	LAK9	2.520
16	−36.2242	0.0004		2.520
17	2.0680	0.5136	LAK9	2.520
18	1.6122	2.3729		2.100
19	0.0000	0.0000		2.246

The distance from the first lens surface to the images 12.759. Distortion = 14%.

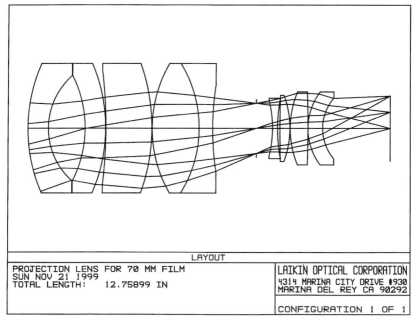

(a)

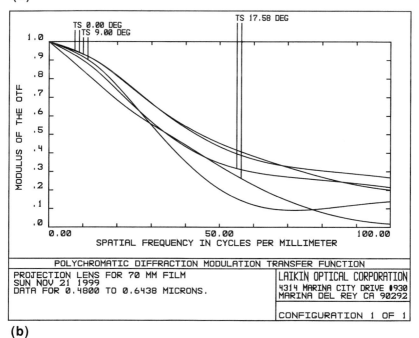

(b)

Figure 22-3 (a) f/2 lens. (b) Lens MTF.

The entrance pupil is 4.747 from the first lens surface. The exit pupil is −3.974 from the image (film) surface. Note that the diameters of the front lenses were made the same, as this simplifies the lens housing design.

Figure 22-4 shows a 70° FOV, f/2 projection lens for 70 mm film. EFL = 1.5.

Table 22-4 70° FOV Projection Lens

	Radius	Thickness	Material	Diameter
0	0.00000	1.00E+10	Air	
1	10.59621	0.5398	SK14	5.480
2	0.00000	0.6701	Air	5.480
3	5.67687	0.2988	SF2	3.320
4	1.28858	1.3654	Air	2.280
5	−22.98210	0.4491	FK5	2.000
6	1.28091	0.2279	Air	1.700
7	3.32654	0.4939	SF6	1.780
8	−9.84090	0.5778	Air	1.780
9	Stop	0.1031	Air	1.552
10	3.04039	0.6438	LAF2	1.900
11	−2.22659	0.0240	Air	1.900
12	26.35518	0.5294	FK5	1.860
13	−1.31642	0.1060	SF6	1.860
14	2.32165	0.1862	Air	1.860
15	33.26503	0.3124	LAK9	1.900
16	−2.75162	0.0238	Air	1.900
17	5.26397	0.2939	LAF9	2.100
18	−8.10888	2.2448	Air	2.100

The distance from the front lens surface to the image is 9.090. Distortion at full field is 4.6%. The entrance pupil is 2.944 from the first lens surface. The exit pupil is −4.522 from the film. Referring to the MTF plot in Fig. 22-4b, you will note the excellent response on axis. At 35° off axis, there is a 10% response for the tangential component and a 17% response for the saggital component. Assuming then a

Projection Lenses

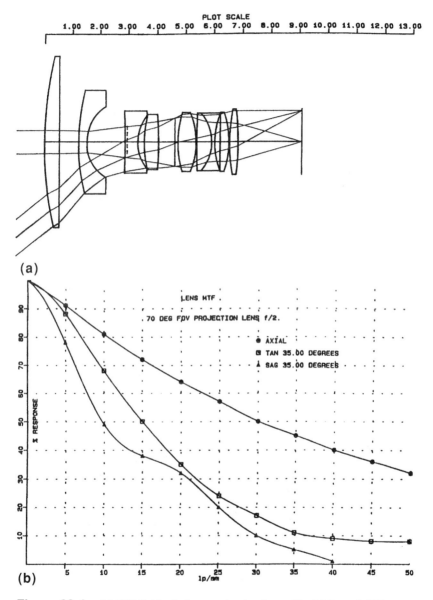

Figure 22-4 (a) 70° field of view projection lens, f/2. (b) Lens MTF.

30 lp/mm resolution, someone in the rear of the theater would see an angular resolution at the corner of the screen of

$$\theta = \frac{1}{30(38.1)} = 0.00087 = 3 \text{ minutes of arc}$$

As discussed above, a patron in the center of the theater would have an angular resolution of 6 minutes of arc at the corner of the screen. This is adequate for most applications.

Figure 22-5 shows a plastic projection lens (10 inch focal length, f/2) designed to project images from a 5 inch diameter CRT tube. Plastic was chosen because the client wanted a lens that could be injection molded for high volume, low-cost production. Since there are relatively low power densities here, the use of plastics is not precluded.

The aperture stop is located between the second and third lenses. This is simply the spacer between these two lenses. Distortion is 1.9%.

Table 22-5 Plastic Projection Lens

	Radius	Thickness	Material	Diameter
0	0.00000	1.00000E+08	Air	
1	6.61320	0.8152	Acrylic	5.720
2	−98.64314	0.0183	Air	5.720
3	2.87612	1.2556	Acrylic	4.880
4	8.09377	0.2322	Air	4.420
5	27.63460	0.3986	Polystyrene	4.420
6	2.39872	0.5505	Air	3.780
7	4.90627	0.3937	Polystyrene	3.940
8	3.74047	1.5837	Air	3.840
9	5.12128	1.2149	Acrylic	5.590
10	−15.37367	4.6860	Air	5.590

The distance from the first lens surface to the image is 11.149.

The acrylic material is polymethyl methacrylate and is listed in the glass file described in Appendix E to be used with the Refractive

Projection Lenses

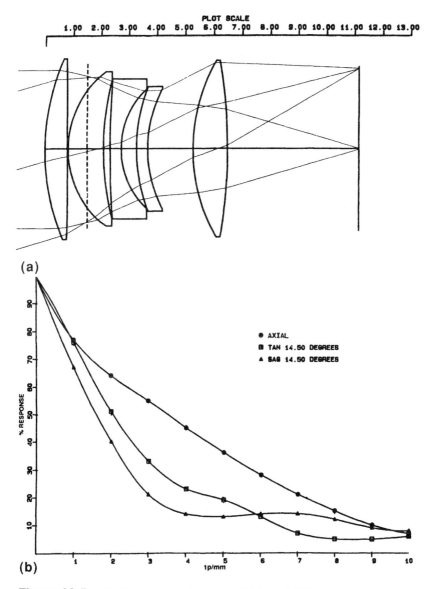

Figure 22-5 (a) Plastic projection lens. (b) Lens MTF.

Index program, as Plexiglas. Both Plexiglas and polystyrene are thermosetting, they are the most common materials used in injection molding. As discussed earlier under aspheric surfaces, this type of injection-molded lens is an ideal candidate for aspheric surfaces.

Figure 22-6 shows an arrangement to project a liquid crystal display (LCD). The dielectric coatings selectively reflect (and transmit the complement) red, green, and blue images from the liquid crystal. Due to the properties of this liquid crystal, the lens should be nearly telecentric.

This lens is shown in Fig. 22-7. The focal length is 2.0 inch, and it is f/5. It has 6.1% distortion. It is a modification of Taylor (1980). It projects an image of 1.75 diameter. The length from the first lens surface to the image is 9.934 inch.

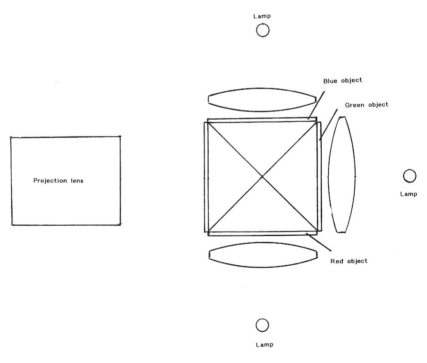

Figure 22-6 LCD projection.

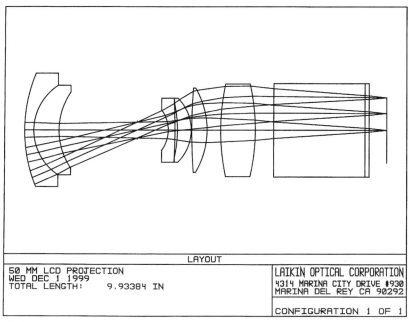

(a)

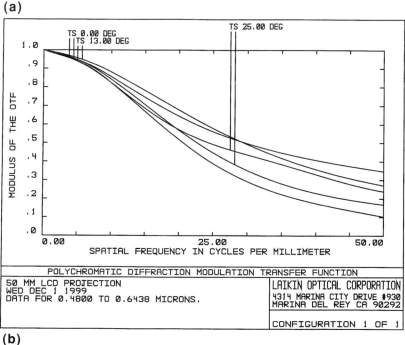

(b)

Figure 22-7 (a) 2.0-in., $f/5$ lens. (b) Lens MTF.

Table 22-7 Lens for LCD Projection

Surf	Radius	Thickness	Material	Diameter
0	0.0000	1.00E+10		
1	3.9304	0.2500	PSK3	3.040
2	1.4231	0.6829	SF4	2.400
3	1.7647	2.1545		2.040
4	Stop	0.8588		0.424
5	−0.8867	0.1874	SF8	1.080
6	−12.8800	0.0661		1.720
7	−3.3142	0.3809	PSK53A	1.470
8	−1.1853	0.0175		1.720
9	14.7759	0.4193	PSK53A	2.260
10	−2.1747	0.3699		2.260
11	5.7671	0.9422	PSK3	2.480
12	−5.4309	0.5000		2.480
13	0.0000	2.5000	BK7	2.500
14	0.0000	0.1127	2KN7	2.500
15	0.0000	0.4916		2.500
16	0.0000	0.0000		1.751

REFERENCES

ANSI PH 22.28 (1982). Dimensions for 35 and 70 mm projection lenses, American National Standards Institute, 1430 Broadway, New York, NY.

ANSI 196M (1986). Screen illuminance and viewing conditions, American National Standards Institute, 1430 Broadway, New York, NY.

Betensky, B. (1982). Projection lens, U.S. Patent #4348081.

Buchroeder, R. A. (1978). Fisheye projection lens system for 35 mm motion pictures, U.S. Patent #4070098.

Clarke, J. A. (1988). Current trends in optics for a projection TV, *Optical Engineering, 27:* 16.

Corbin, R. M. (1969). *Motion Picture Equipment, Applied Optics and Optical Engineering, 5:*305 (R. Kingslake, ed.), Academic Press, New York.

Mittal, M.K. and Gupta, B. (1994). Six element objective based on a new configuration, *Optical Engineering, 33:*1925.

Sharma, K. D. (1982). Better lenses for 35 mm projection, *Applied Optics, 21:*4443.

Sharma, K. D. (1983). Future lenses for 16 mm projection, *Applied Optics, 22:*1188.

Sharma, K. D. and Kumar, M. (1986). New lens for 35 mm cinematograph projector, *Applied Optics, 25:*4609.

Schneider Optics (2000). Schneider Theater Design Pro 2.1b, WWW.Schneideroptics.com.

Taylor, W. H. (1980). Wide angle telecentric projection lens, U.S. Patent #4189211.

Wheeler, L. J. (1969). *Principles of Cinematography,* Fountain Press, Watford, Endland, Chap. 10.

23
Telecentric Systems

A telecentric optical system is one that has its exit pupil at infinity. (The chief ray emerges parallel to the optical axis.) An example of use is in a contour projector where one wants to measure screw threads accurately. Since the chief ray is parallel to the optical axis, measurements are the same regardless of the image location.

The computer program should have the ability to control the exit pupil distance from the last lens surface. However, for the telecentric feature, it is preferable instead to bound the reciprocal of the exit pupil distance. This is because most bounded items are small numbers: lens thickness, refractive index, paraxial height ratios, etc. Bounding the pupil to an infinitely large number would force the program to obey this bound to the exclusion of all the other bounds. Reciprocal pupil distance solves this problem and thus keeps all the bounds within the same magnitude.

Figure 23-1 shows a telecentric f/2.8 lens used as a 20× profile projection lens.

The distance from the object (projection screen) to the first lens surface is 50. The distance from the first lens surface to the image is 9.535. Distortion is 0.5%. (Note that MTF plots are for 95% of full field.) The image diameter is 1.0.

Several mirrors are placed between the lens and the screen to make a compact optical system. An erect image is obtained by the addition of a relay or prisms in conjunction with the mirrors.

Figure 23-2 shows a variation of this lens; the object is now at in-

265

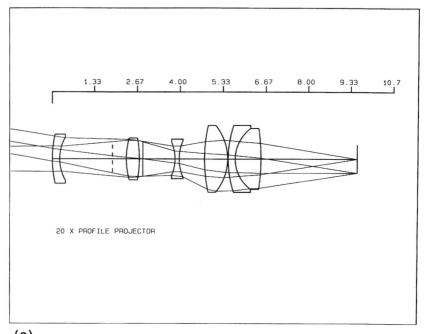

(a)

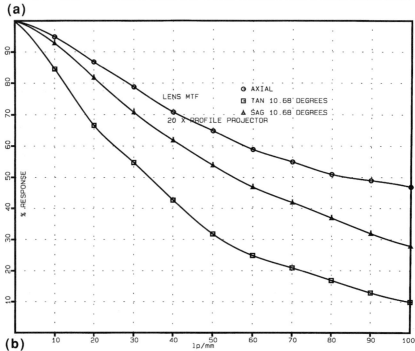

(b)

Figure 23-1 (a) 20× profile projector. (b) Lens MTF.

Table 23-1 20× Profile Projection Lens

	Radius	Thickness	Material	Diameter
0	0.00000	50.000		
1	3.88595	0.2304	K10	1.640
2	1.53641	2.0943	Air	1.440
3	2.82407	0.4079	LAKN12	1.400
4	−4.02049	0.1097	Air	1.400
5	Stop	1.0022	Air	1.168
6	−1.50892	0.1290	F5	1.120
7	1.63110	0.7910	Air	1.220
8	5.72732	0.7130	SK5	2.240
9	−1.88036	0.0182	Air	2.240
10	3.913529	0.2144	SF1	2.240
11	1.33230	0.8045	BALF4	2.040
12	−7.18973	3.0203	Air	2.040

Table 23-2 Telecentric $f/2.8$ Lens

	Radius	Thickness	Material	Diameter
0	0.00000	1.000E+07	Air	
1	−47.63383	0.2638	FK5	1.530
2	2.13147	0.1461	Air	1.450
3	Stop	2.1542	Air	1.446
4	26.39803	0.3942	LAK8	3.340
5	−6.03104	1.7429	Air	3.340
6	5.61254	0.6345	LAK8	3.940
7	−13.21876	0.2750	Air	3.940
8	13.78651	0.3389	SF1	3.660
9	3.06698	0.3894	Air	3.330
10	11.76266	0.2807	SF1	3.400
11	3.04728	0.8596	LAKN12	3.400
12	−5.79673	5.2507	Air	3.400

The Distance from the first lens surface to the image is 12.730.

finity. The aperture stop is immediately behind the first lens. EFL = 4. Field of view = 20 degrees. Distortion is 0.89%. For economy in production, the first lens surface should be set plane.

Figure 23-3 shows a f/2 40 degree field of view telecentric lens. The effective focal length is 1.0. The distance from the first lens surface to the image is 8.062. Distortion = 1.3%.

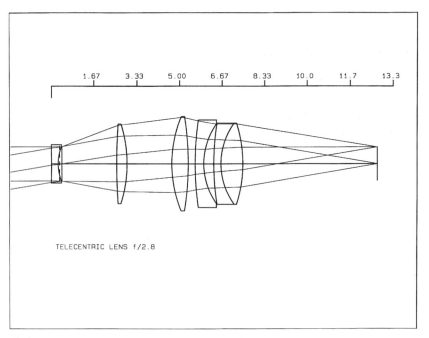

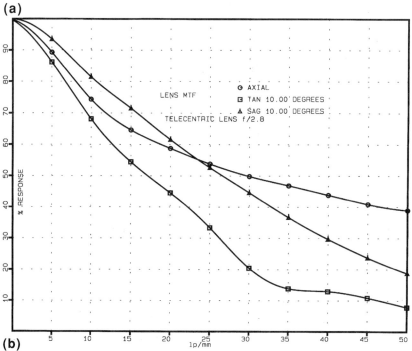

Figure 23-2 (a) Telecentric *f*/2.8 lens. (b) Lens MTF.

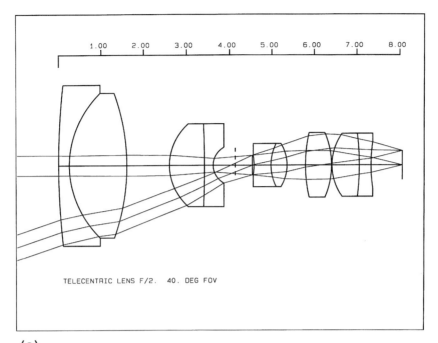

(a)

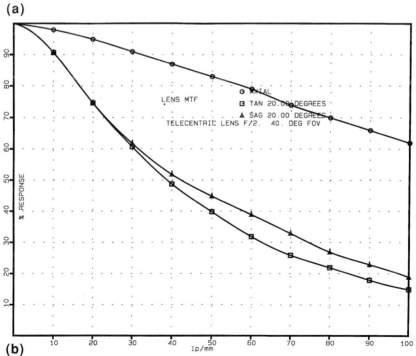

(b)

Figure 23-3 40° field of view, $f/2$ telecentric lens. (b) Lens MTF.

Table 23-3 20° FOV Telectric Lens

	Radius	Thickness	Material	Diameter
0	0.00000	1.00000E+07	Air	
1	19.61108	0.2495	LAFN21	4.020
2	2.61682	1.3560	BK7	3.620
3	−5.66570	0.9991	Air	3.620
4	1.45549	0.8142	SF4	2.080
5	−25.73312	0.2006	BAFN10	2.080
6	0.53295	0.9160	Air	0.900
7	Stop	0.0418	Air	0.490
8	−1.55613	0.3968	SF4	0.520
9	1.26921	0.3767	SK16	1.080
10	−1.08945	0.4385	Air	1.080
11	3.94797	0.6082	LAKN12	1.620
12	−2.17713	0.0108	Air	1.620
13	1.41656	0.6481	SK16	1.580
14	−6.95994	0.2668	SF4	1.580
15	4.20851	0.7388	Air	1.160

REFERENCES

Dilworth, D. (1971). Telecentric lens system, U.S. Patent #3565511.

Reiss, M. (1952). Telecentric objective of the reverse telephoto type, U.S. Patent #2600805.

Tateoka, M. (1984). Telecentric projection lenses, U.S. Patent #4441792.

Young, A. W. (1967). Optical workshop instruments, *Applied Optics and Optical Engineering, 4:*250, Academic Press, New York.

Zverev, V. A. and Shagal, A. M. (1976). Three component objective lens, *Sov. J. Opt. Tech., 43:*529.

24
Laser-Focusing Lenses

Lenses for video and optical disk use are small, high-numeric-aperture lenses operating at a single laser wavelength. They cover a very small field, are diffraction limited, and are nearly aplanatic.

Figure 24-1 shows an f/1 lens for video disc use for a HeNe laser (0.6328 μm). The field of view is 1 degree. EFL = 02. This design is based on the Minoura (1979) patent.

Table 24-1 Video Disk Lens

	Radius	Thickness	Material	Diameter
0	0.00000	1.000E+06	Air	
1	0.46045	0.0747	SF6	0.250
2	−1.19418	0.0238	Air	0.250
3	−0.27590	0.0225	BK7	0.190
4	−0.57779	0.0693	Air	0.250
5	0.14885	0.1094	SF6	0.190
6	0.59850	0.0838	Air	0.190

The distance from the first lens surface to the image is 0.384. Distortion = 0.3% (pincushion). The entrance pupil is located 0.072 from the first lens surface.

A recent trend in the manufacture of these lenses is to injection mold in plastic or in glass (Fitch, 1991) This has the advantages of light weight and low cost, and an aspheric surface can be used. Another technique is to replicate a thin layer of plastic onto a molded glass lens (Saft, 1994).

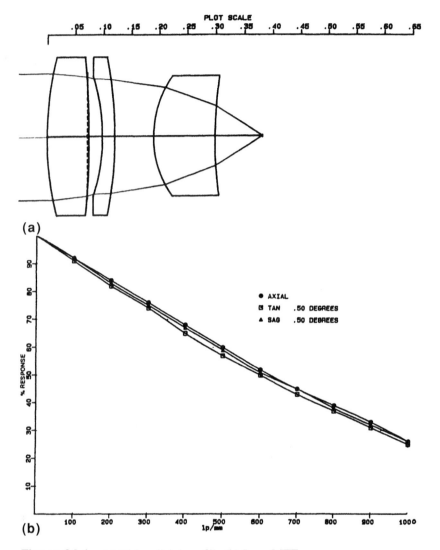

Figure 24-1 (a) Video disk lens f/1. (b) Lens MTF.

Laser-Focusing Lenses

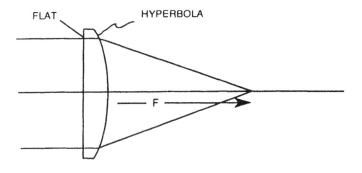

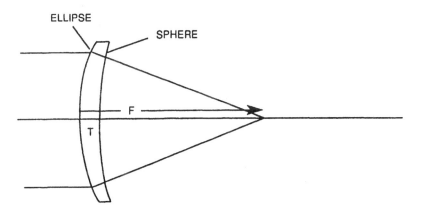

Figure 24-2 Conic section lenses.

It is true that mold costs are very high, but once these are made lens cost in large volume is very low. It has the added advantage that the spacers and mounting means are molded into the lens.

Using conic sections, it is possible to have a single lens with zero spherical aberration. Referring to Fig. 24-2 we see that for the flat hyperbola case; a plane wavefront is incident on the flat and is refracted at the hyperbolic surface so that a spherical wavefront emerges. If the lens has an index of refraction N, R is the paraxial radius, and ε is the eccentricity ($A_2 = -\varepsilon^2$ as used in Grey's POP and the ZEMAX programs; see the section on aspheric surfaces). Thus

$$R = -F(N-1) \quad \varepsilon = N$$

For the ellipse–sphere case we have $F = N + 1$.
For the ellipse,

$$R = \frac{N^2 - 1}{N} \quad \varepsilon = \frac{1}{N}$$

For the sphere, $R = EFL - T$.

Both cases show diffraction-limited performance on axis. Due to excessive coma, performance is very poor even 0.5 degrees off axis. For either of these cases to be effective, the beam would have to be aligned to the lens optical axis to within a few minutes of arc.

Figure 24-3 shows a lens used to focus an excimer laser (XeCl 0.308 μm). It has a focal length of 5 and a 1 diameter entrance pupil. FOV = 1 deg. This correction of coma means that the laser beam does not have to be precisely aligned to the lens optical axis.

Table 24-3 Laser Focusing Lens, 0.308μm

	Radius	Thickness	Material	Diameter
0	0.00000	1.00000E+07	Air	
1	3.17381	0.2592	Silica	1.100
2	−6.10489	0.1395	Air	1.100
3	−2.43609	0.2496	Silica	0.960
4	−2.88562	4.6691	Air	1.100

The distance from the first lens surface to the image is 5.317.

Figure 24-4 shows a laser focusing lens to be used with an HeNe laser (0.6328 micron). It can be used as a bar code reader lens.

Table 24-4 Laser Focusing Lens, 0.6328μm

Surf	Radius	Thickness	Material	Diameter
0	0.0000	1.0000E+10	Air	
1	0.30153	0.2300	Polystyrene	0.500
2	−3.4042	0.3469	Air	0.500

Laser-Focusing Lenses

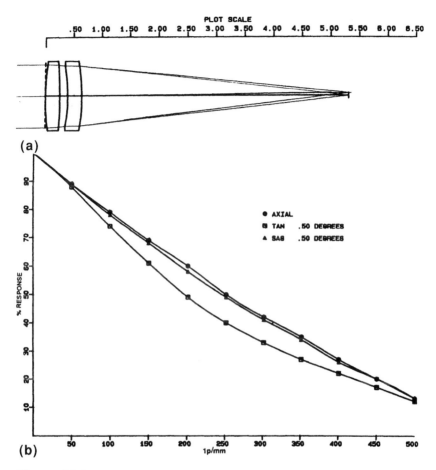

Figure 24-3 (a) Laser-focusing lens 0.308 µm. (b) Lens MTF.

The effective focal length is 0.4833, and it is f/1. It is nearly diffraction limited for a field of view of 1.0 degree. At 0.6328 µm the refractive index of polystyrene is 1.58662. The first surface is aspheric. Its equation is

$$X = \frac{3.316436Y^2}{1+\sqrt{1-7.01341Y^2}} - 1.174228Y^4 - 10.87748Y^6 - 25.67630Y^8 - 721.13841Y^{10}$$

276 *Chapter 24*

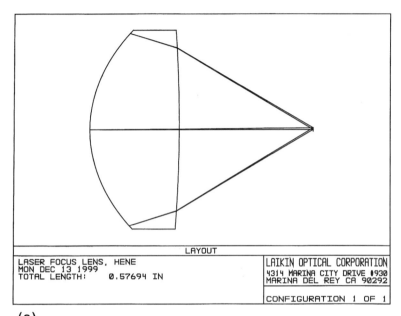

(a)

(b)

Figure 24-4 (a) Laser-focusing lens for an HeNe laser. (b) Lens MTF.

REFERENCES

Binnie, T. D. (1994). Fast imaging micro lenses, *Applied Optics, 33:*1170.
Broome, B. G. (1992). Proceedings of Lens Design Conference, Jan. 1992, Los Angeles, SPIE Press.
Chirra, R. R. (1983). Wide aperture objective lens, U.S. Patent #4368957.
Fitch, M. A. (1991). Molded optics, *Photonics Spectrs.,* Oct., 1991.
Isailovic, J. (1987). *Videodisc Systems,* Prentice Hall, NJ.
Minoura, K. (1979). Lens having high resolving power, U.S. Patent #4139267.
Saft, H. W. (1994). Replicated optics, *Photonics Spectrs.,* Feb., 1994.
U.S. Precision Lens (1973). *The Handbook of Plastic Optics,* 3997 McMann Rd, Cincinnati, Ohio 45245.

25
Heads Up Display

This type of lens is used in the cockpit of an aircraft to image a cathode ray tube into the pilot's eyes. It is like a huge eyepiece, with the exit pupil being at the pilot's eyes. This pupil should be large enough to encompass both eyes and allow for some head movement. Typical specifications for such a system would be

Entrance pupil diameter	6 inches
Pupil distance to lens	25 inches
Angular resolution	
center	1 minute of arc
edge	3 minute of arc
Field of view	25 degrees
Instantaneous field of view	20 degrees
Maximum distortion	5%

Singh (1996) gives additional requirements. For this system, one can assume that there are two 8 mm diameter pupils 65 mm apart. What one eye can see, without moving the head, is referred to as "instantaneous field of view." Due to limitations on the size of the optics, one generally has to move one's head around a little in order to see the full CRT display.

Due to size limitations in the cockpit, an in-line system will not fit. The lens then has to be "folded" to fit into the crowded cockpit. Figure 25-1 shows a typical heads up display (see also Rogers, 1980). Since it was designed to be used with a 4.25 inch diameter CRT and have a field of view of 25 degrees, the focal length becomes 9.585.

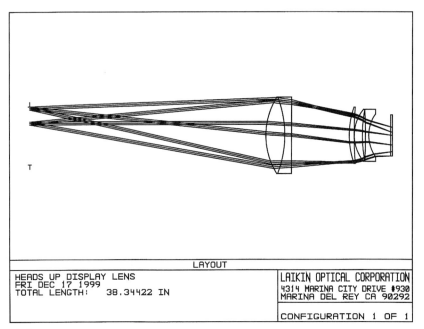

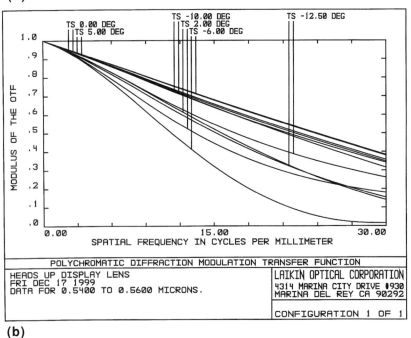

Figure 25.1 (a) Heads up display lens. (b) Lens MTF.

Heads Up Display

Table 25-1a Heads Up Display

Surf	Radius	Thickness	Material	Diameter
0	0.0000	1.0000E+10		
1	Stop	25.0000		6.000
2	7.4702	1.9817	SK16	7.900
3	−10.7389	0.7000	SF6	7.900
4	0.0000	6.1000		7.900
5	8.6772	0.5212	SF6	5.840
6	34.0874	0.0200		5.840
7	4.1812	0.8597	SF6	5.400
8	13.8540	0.3291		5.200
9	−50.2546	0.3500	SF6	5.400
10	3.8134	2.2935		4.440
11	0.0000	0.1890	K5	4.280
12	0.0000	0.0000		4.280
13	0.0000	0.0000		4.267

The distance from the first lens surface to the image is 13.344. The space between the first and second lenses allows the insertion of a mirror to fold the optical system for convenient packaging of the lens into the cockpit. The last element of K5 glass is the CRT face plate. Due to the difficulty of chromatic correction in these types of systems, they are generally designed to be used with a narrow band CRT phosphor (for example, P53 phosphor). In this case, the central wavelength is .55 μm with the half power points being at 0.54 and 0.56 μm. This narrow bandwidth also has another advantage. These systems are generally positioned so that the CRT image is reflected from the aircraft windshield into the pilot's eyes. In order to cause a minimum transmitted energy loss at the windshield, the windshield (or a beam-splitter very close to this windshield) is coated with a narrow band dielectric reflecting coating.

From the trace in Fig. 25-1 notice that the 8 mm diameter pupil at the left eye position can view 5 degrees to the left and 10 degrees to the right. The same is true for the right eye. Thus we say that it has an instantaneous field of view of 20 degrees. By moving the head within a 6 inch diameter circle, the full 25 degree field will be visible. Distortion = 0.2%.

Table 25-1b Pupil Shift vs. Field Angle for a Heads Up Display

Field	Pupil shift (from center)	Angle (deg)
1	2.842	2.0
2	2.842	–6.0
3	2.842	–12.5
4	1.279	0.0
5	1.279	5.0
6	1.279	–10.0
7	1.279	2.5 sagittal
8	1.279	5.0 sagittal

Since the pupil is displaced, not only must the computer program have the ability to displace the entrance pupil, but it must also be able to trace positive and negative field angles in addition to field angles in the sagittal plane. In the above example, nine field angles were used, seven in the tangential plane and two in the sagittal plane. In Table 25-1b these field angles are given.

Resolution in the central portion of the field is one minute of arc. At the edge, resolution is reduced due to astigmatism. Field angles 4 and 5 are indicated on the MTF plot.

A biocular lens is similar in many respects to a heads up display. However, here the user's eyes are much closer to the lens than in a heads up display. They are used in various types of viewing equipment where a relatively low power eyepiece is required, and the convenience of viewing through both eyes is desired. As with a heads up display, the computer program must be able to shift the entrance pupil and trace from objects in both tangential and sagittal planes. Figure 25-2 shows a biocular lens of focal length 4. From the formula for magnification in Chap. 10,

$$M = \frac{4+10}{4} = 3.5$$

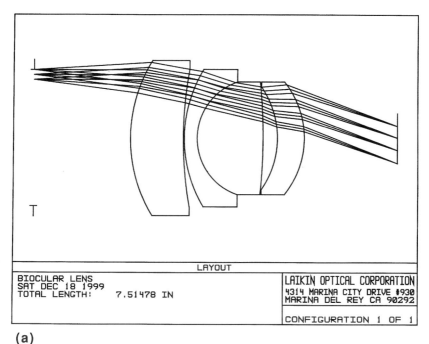

(a)

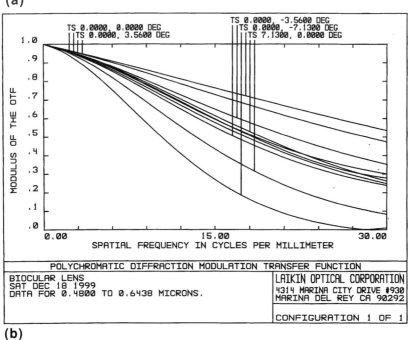

(b)

Figure 25.2 (a) Biocular lens of focal length 4. (b) Lens MTF.

This lens has visual correction and covers an object diameter of 1. Rays were traced for a 5 mm pupil diameter, 32.5 mm from the lens center line.

Table 25-2a Biocular Lens

Surf	Radius	Thickness	Material	Diameter
0	0.0000	1.0000E+10		
1	Pupil	2.0000		2.757
2	2.9339	1.1001	PSK3	3.140
3	7.7247	0.0200		2.800
4	2.6343	0.2680	SF2	2.780
5	1.1996	1.3501	FK5	2.280
6	−10.7987	0.2943		2.280
7	−1.8077	0.5547	LF5	2.080
8	−1.7849	1.9276		2.280

The distance from the first lens surface to the image is 5.515.

It is interesting in that resolution off axis is much better than the axial resolution. The reason for this can be seen from the ray trace. Note that the rays from an axial object travel near the edge of the lens, whereas the rays from an object at the edge of the field are passing nearly in the center of the lens.

Table 25-2b Field Angles for Biocular Lens

Field	Angle (deg)	Displacement	Distortion (%)
1	0.0	1.28	0.0
2	3.56	1.28	0.8
3	−3.56	1.28	2.3
4	−7.13	1.28	2.0
5	7.13 sagittal	1.28	1.8

REFERENCES

JEDEC Pub. #16-C (1988). Optical characteristics of cathode ray tubes, Electronic Industries Association, 2001 Eye St. NW, Washington, DC 20006.

Rogers, P. J. (1980). Modified Petzval lens, U.S. Patent #4232943.

Singh, I, Kumar, A, Singh, H. S., and Nijhawan, O. P. (1996). Optical design and performnce evaluation of a dual beam combiner HUD, *Optical Engineering, 35:*813.

26
Achromatic Wedges

The achromatic wedge is a useful device to obtain a small angular deviation and still preserve an achromatic image. It is limited to small deviation angles for two reasons:

1. Individual prism angles become excessively large for large deviation angles.
2. Secondary color increases with deviation.

To prevent the introduction of coma and astigmatism into the axial image (see Chap. 27), it is important that these wedges be used only in a collimated light bundle. Sometimes (unfortunately) prism deviation is measured by ophthalmic nomenclature:

$$1 \text{ prism diopter} = \frac{1 \text{ cm deviation}}{1 \text{ m distance}}$$

Figure 26-1 shows an incident beam normal to the first prism surface. After refraction, the beam emerges with a deviation as shown. Smith (1966) gives useful first-order formulas for calculating the prism angles. The author wrote a program that traces a real ray through the prism assembly and, by a simple iterative scheme, calculates the prism angles for an achromatic condition and to obtain the desired deviation. (The deviation is calculated at the central wavelength, and to be achromatic, the deviation is to be the same at the extreme wavelengths.)

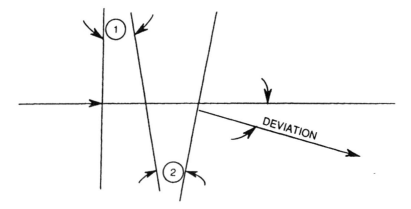

Figure 26-1 Achromatic prism.

Table 26-1 Achromatic Prism

Deviation (deg)	Angle 1 (deg)	Angle 2 (deg)	Secondary (microradians)
1	4.419	2.071	10
2	8.781	4.098	19
3	13.031	6.040	29
4	17.127	7.865	38
5	21.032	9.546	48
6	24.719	11.070	58
7	28.176	12.430	68
8	31.393	13.625	78
9	34.374	14.663	89
10	37.124	15.554	99

Consider an achromatic prism in the visual region. BK7 glass is used for the first prism and F2 for the second. See Table 26-1.

The table was computed using the Achromatic Prism program in Appendix E. Secondary is the angular difference between the deviation at the central wavelength and the deviation at the short wavelength.

REFERENCES

Goncharenko, E. N. and Repinskii, G. N. (1975). Design of achromatic wedges, *Sov. J. Opt. Tech., 42:*445.

Sheinus, N. V. (1976). Design of a wedge scanner, *Sov. J. Opt. Tech., 43:*473.

Smith, W. (1966). *Modern Optical Engineering,* McGraw-Hill, New York, Sec. 4.5.

27
Wedge Plates and Rotary Prisms

Quite frequently it is necessary to place a beam-splitter in a convergent beam. If a plate is used, it will introduce coma and astigmatism into the axial image. This can be reduced or eliminated by

1. Using a pellicle. This is equivalent to a plate of zero thickness. Pellicles are made by forming a nitrocellulose plastic over a flat frame. They are extremely thin, typically about 5 μm. Limited types of coatings may be applied to this rather fragile part. Glass pellicles of perhaps 0.005 inch thick glass are also available. Although thicker than the plastic form, they are more durable, and a large variety of coatings may be applied.

2. Using a cube beam-splitter. This is equivalent to a plane plate in the system and is easily allowed for in the optical design. They are bulky and sometimes there is no way to incorporate them into the mechanical package. In the IR region, a cube is out of the question.

3. Using a wedge plate. By this technique axial coma and astigmatism can be substantially reduced.

Referring to Fig. 27-1 we see a plate of thickness T with a wedge angle of θ in a converging beam. It is a distance P to the image. The plate is at an angle β to a normal to the optical axis.

For zero astigmatism, De Lang gives the following value for the wedge angle:

$$\theta = \frac{\sin\beta(\cos^2\beta)T}{2(N^2 - \sin^2\beta)P}\text{ rad}$$

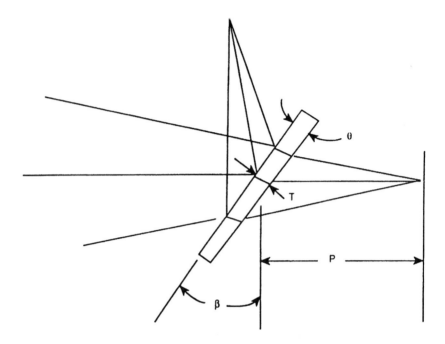

Figure 27-1 Wedge plate.

Coma is also substantially reduced.

I once had to incorporate a beam-splitter into a 10× microscope objective system. The beam splitter had the following parameters:

$N = 1.51872$ (BK7 glass) $B = 45$ degrees
$T = 0.04$ inch $P = 14.9473$ inch

Using these values in the above equation we obtain for θ, the wedge plate angle, 0.00026186 radian = 54.0 seconds.

Figure 27-2 shows the results of this by means of spot diagrams of the axial image. For comparison, a spot diagram is also presented for a parallel plate.

Wedge Plates and Rotary Prisms

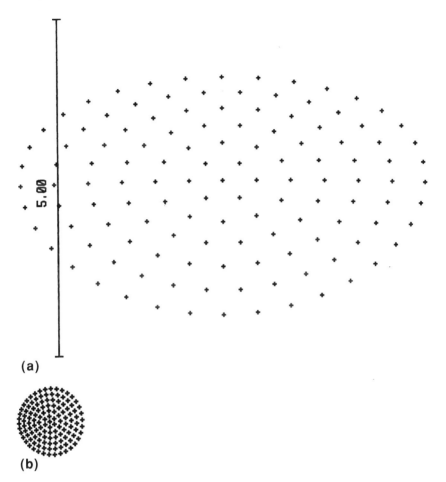

Figure 27-2 (a) Parallel plate at 45°. (b) Wedge plate at 45°.

4. Using two beam-splitters oriented 90 degrees to each other. That is, if we look along the optical axis, we see two reflected beams at a 90 degree orientation to each other.

A plane plate can be used to displace a beam and is the basis of rotary prism cameras. In this type of camera, a plate is coupled to the film movement so as to cause a displacement equal to the film move-

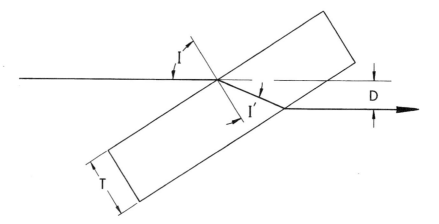

Figure 27-3 Ray displacement due to a plate.

Table 27-4 Lens for Rotary Prism Camera

Surf	Radius	Thickness	Material	Diameter
0	0.0000	1.0000E+10		0.00
1	0.6772	0.3057	SF1	1.000
2	0.4532	0.1514		0.700
3	0.8751	0.2287	LAF3	0.740
4	−2.8782	0.0704	F2	0.740
5	0.7460	0.0845		0.560
6	Stop	0.3159		0.532
7	−0.8706	0.1083	F2	0.720
8	2.7059	0.2949	N-LASF30	1.000
9	−0.8075	0.2816		1.000
10	1.2784	0.3380	LAKN12	1.040
11	−0.9851	0.1408	SF1	1.040
12	21.6185	0.3300		0.880
13	0.0000	0.4480	LAK16A	0.448
14	0.0000	0.3300		0.448
15	0.0000	0.0000		0.496

Distortion is 2.2%. The distance from the first lens surface to the image is 3.428.

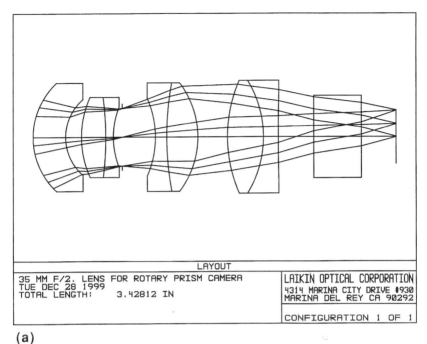

(a)

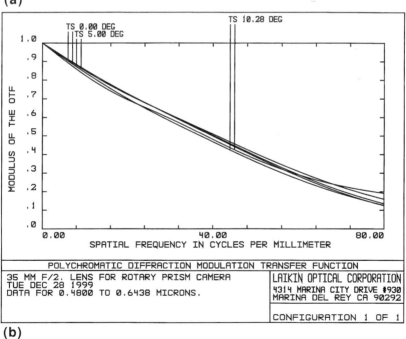

(b)

Figure 27-4 (a) 35 mm, f/2 lens for a rotary prism camera. (b) Lens MTF.

ment. The problem, in addition to aberrations of the plate, is that this displacement is not a linear function of the plate rotation. Figure 27-3 shows the displacement of a plane plate of thickness T and refractive index N. We have

$$\sin I = \frac{\sin I'}{N} \qquad D = \frac{T\sin(I - I')}{\cos I'}$$

Due to the nonlinear displacement of the plate, a shutter is provided in rotary prism cameras to cut off the beam to limit this displacement error.

In Fig. 27-4 is shown a lens for a rotary prism camera. To reduce the effects of astigmatism as the plate is rotated, the lens is made nearly telecentric (Buckroeder, 1977).

The local length is 1.378 (35 mm), it is f/2, and it will cover the 16 mm film format (see Appendix A).

REFERENCES

Buckroeder, R. A. (1977). Rotating prism compensators, *J. SMPTE, 86:*431.

Howard, J. W. (1985). Formulas for the coma and astigmatism of wedge prisms in convergent light, *Applied Optics, 24:*4265.

DeLang, H. (1957). Compensation of aberrations caused by oblique plane parallel plates, *Philips Research Reports, 12:*131.

Sachteben, L. T., Parker, D. T., Allee, G. L., and Kornstein, E. (1952). Color television camera system, *RCA Review, 8:*27.

Shoberg, R. D. (1978). High speed movie camera, U.S. Patent #4131343.

Whitley, E. M., Boyd, A. K., and Larsen, E. J. (1996). High speed motion picture camera, U.S. Patent #3259448.

28
Anamorphic Attachments

A recent trend in cinematography is to provide the audience with a wide field of view. This of course can be accomplished by using a wider film (70 mm rather than 35 mm). However, it is more economical and certainly more convenient to be able to photograph and project using 35 mm film. Chretien (1931) was perhaps the first to devise a process using anamorphic lenses.

Traditional 35 mm projection has a horizontal width-to-height ratio of 1.37. In the CinemaScope process, a 2× anamorphic afocal attachment is placed in front of the prime lens. This attachment has no effect in the vertical dimension but compresses the horizontal image on the film. Although standard 35 mm film is used, the vertical height on the film is slightly larger than standard (unsqueezed) cinematography. The net result is an aspect ratio of 2.35 (Wheeler, 1969, p. 76).

Projection is accomplished in the same manner: an afocal cylinder attachment is placed in front of a standard projection lens. This expands the horizontal field by a factor of two.

In Fig. 28-1 such a projection attachment is shown. It was meant to be used with a 3 inch f/2 projection lens for 35 mm cinematography. This is an adaptation from Cook (1958) (view is in the cylinder axis power plane).

The distance from vertex to vertex is 9.903. This lens is a little too long; the large air space should be reduced. Distortion in the tangential plane is 1.8%.

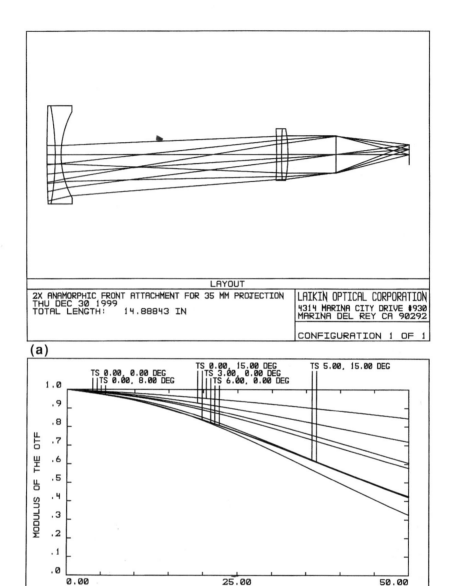

Figure 28-1 (a) 2× anamorphic front attachment for 35 mm projection. (b) Lens MTF.

Table 28-1 2× Anamorphic Front Attachment

Surf	Radius	Thickness	Material	Diameter
0	0.0000	1.0000E+10		0.00
1	46.7733	0.3497	SF5	3.940
2	−10.3251	0.2416	BK10	3.940
3	3.5659	8.8451		3.460
4	−19.2836	0.2006	BASF2	2.040
5	21.4963	0.2658	K5	2.080
6	−5.5945	1.9830		2.080
7	Exit pupil	3.0025		2.001
8	0.0000	0.0000		0.943

In Fig. 28-2 is shown a 2× anamorphic attachment for use with 70 mm film and a 4 inch focal length, f/2 projection lens (view is in the cylinder axis power plane).

The first surface should be set plane for production. The distance from vertex to vertex is 10.278. Distortion, in the tangential plane, is 0.5%.

As with the above case, the MTF axial values plotted are for the tangential orientation. Axial sagittal values are the same as off axis (object still in the tangential plane).

These attachments are made so as to permit adjusting the air space for the actual lens-to-screen distance (usually 50 to 400 feet). Since this is generally a large distance, this adjustment creates no problem. However, in the studio, at short close-up distances, the anamorphic ratio substantially changes as well as introducing astigmatism. This has the effect in projection of making the actors appear fatter than they actually are. This proved to be a fatal flaw in the original CinemaScope system. This was solved by Wallen and is now the basis of Panavision systems. Wallen (1959) introduced two weak cylinder lenses that are rotated in opposite directions as the attachment is focused. (Actually, the patent discusses astigmatism reduction.) Another method, but one not as convenient, is to place the anamorphic device in convergent light between the prime lens and the image (Kingslake, 1960). The prime lens is moved to focus on the object. Since the rear anamorphic device does not move, the anamorphic magnification remains constant.

(a)

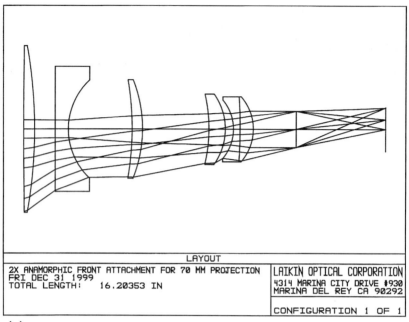

(b)

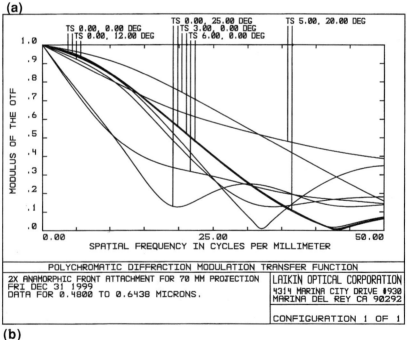

Figure 28-2 (a) 2× anamorphic front attachment for 70 mm projection. (b) Lens MTF.

Table 28-2 2× Anamorphic for 70mm

Surf	Radius	Thickness	Material	Diameter
0	0.0000	1.0000E+10		
1	245.3505	0.5063	SK14	7.280
2	−19.3729	0.9290		7.280
3	0.0000	0.5930	SK14	5.460
4	2.8590	2.8612		4.280
5	−12.2116	0.4711	SF1	4.200
6	−5.6140	2.9392		4.300
7	−6.5068	0.7266	K10	3.020
8	−2.6171	0.3618		3.080
9	−1.9681	0.2990	LAF3	2.560
10	−20.8007	0.5907	FK5	2.840
11	−2.1197	1.9320		2.840
12	Exit pupil	3.9936		1.588
13	0.0000	0.0000		1.923

A system of this type, a rear expander, is shown in Fig. 28-3.

The distance from the first lens surface to the image is 3.731. It was designed to be used with an f/4 camera lens. The camera lens image should fall 3.325 beyond the first lens surface. Distortion at the top of the image (the tangential plane) is 0.4%. The lens is plotted in the power plane (tangential plane). Note that this is a 1.5× expansion, not the usual 2× as in commercial cinematography.

Of course, one can also create anamorphic results by means of prisms (Newcomer, 1933. However, these are awkward to mount and the beam is displaced from the lens axis. In Fig. 28-4 is shown such an anamorphic prism assembly. It consists of two identical cemented prisms. It was designed using the Achromatic Prism program in Appendix E.

Prism angle A = 37.1246 degrees
Prism angle B = 15.554 degrees

The first prism assembly deviates the beam 10 degrees. The second prism has its first surface tilted 10 degrees.

As can seen from the figure, the beam is displaced by 0.4991. For an input beam diameter of 2.0, the output beam diameter is 1.7367, giving a magnification of 1.15.

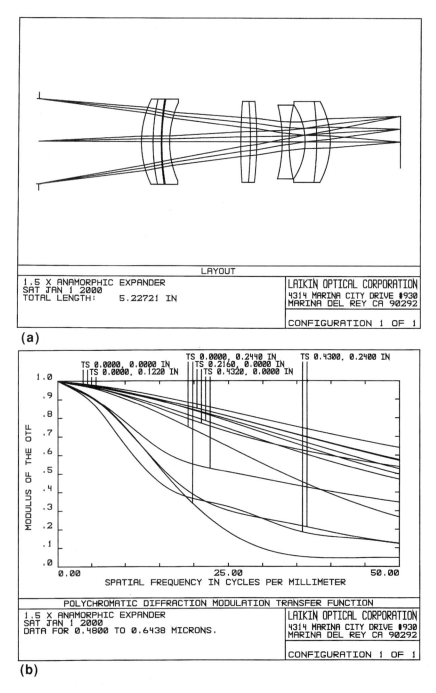

Figure 28-3 (a) 1.5× anamorphic expander. (b) Lens MTF.

Anamorphic Attachments

Table 28-3 Rear Expander

Surf	Radius	Thickness	Material	Diameter
0	0.0000	−4.8210		0.985
1	Entrance pupil	1.4959		1.200
2	1.3427	0.1692	FK5	1.200
3	2.6937	0.1180	SF1	1.200
4	3.9371	0.0100		1.200
5	3.4446	0.1111	LAF2	1.200
6	1.3284	1.0172		1.100
7	8.0086	0.1385	FK5	1.140
8	−6.5103	0.0924	SF1	1.140
9	−5.3699	0.3703		1.140
10	−2.1393	0.0999	FK5	1.040
11	2.2439	0.2314		1.040
12	−0.9081	0.1671	PSK53A	1.040
13	−4.2869	0.1955	LF5	1.140
14	−1.1707	1.0106		1.140
15	0.0000	0.0000		1.124

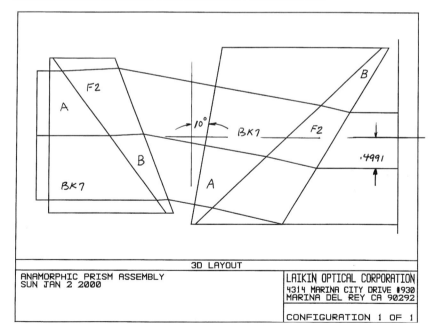

Figure 28-4 Anamorphic prism assembly.

REFERENCES

American Cinematographer Manual (1993). American Society of Cinematographers, 1782 N Orange Dr, Hollywood, CA 90028.

Betensky, E. I. (1977). Continuously variable anamorphic Lens, U.S. Patent #4017160.

Chretien, H. (1931). Process for taking or projecting photographic or cinematographic panoramic views, U.S. Patent #1829633.

Cook, G. H. (1958), Anamorphic attachments for optical systems, U.S. Patent #2821110.

Kingslake, R. (1960). Anamorphic lens system for use in convergent light, U.S. Patent # 2933017.

Larraburu, P. M. (1972). Anamorphic lens systems, U.S. Patent #3644037.

Newcomer, H. S. (1933). Anamorphising prism objectives, U.S. Patent #1931992.

Powell, J. (1983). Variable anamorphic lens, *Applied Optics, 22:*3249.

Raitiere, L. P. (1958). Wide angle optical system, U.S. Patent #2822727.

Rosin, S. (1960). Anamorphic lens system, U.S. Patent #2944464.

Schafter, P. (1961). Anamorphic attachment, U.S. Patent #3002427.

Vetter, R. H. (1972). Focusing anamorphic optical system, U.S. Patent #3682533.

Wallin, W. (1959). Anamorphosing system, U.S. Patent #2890622.

Wheeler, L. J. (1969). *Principles of Cinematography,* Fountain Press, Argus Books, 14 St James Rd, Watford, Herts, England.

29
Illumination Systems

A condensing lens transfers energy from the light source to the entrance pupil of the projection lens. It must also provide uniform illumination at the film (or other object that is to be illuminated). The assumption of a circular uniformly illuminated pupil is imbedded in the usual illumination formulas given in texts. See Smith, 1966, Eq. 8.11. This can be rewritten as

$$E = \frac{T\pi B}{4\left[f^{\#}(m+1)\right]^2} = \text{Illuminance produced at the image}$$

where T is the system transmission, B the object brightness (luminance), $f\#$ the f number of the projection lens, and m its absolute value of magnification. If A is the area of the film, then for very large projection distances ($M \gg 1$),

$$\text{Flux at the screen} = \frac{T\pi BA}{4\left(f^{\#}\right)^2} \text{ lumens}$$

Therefore, for very large screen systems, the tendency is to use a large format film. (See Appendix A for the various film formats.)

Although the usual lens optimization program was designed to minimize image aberrations, such a program can still be used in the design of a refractive condenser system. Several points to remember are

1. Locate the aperture at the film plane.
2. Provide a system magnification such that the source will fill the entrance pupil of the projection lens.

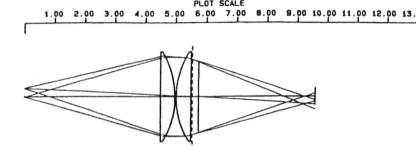

Figure 29-1 Fused quartz condenser 1×.

3. Generally no vignetting.
4. Adjust the program for large emphasis on boundary violations and little on image errors.
5. Due to the generally small source, two field angles should be adequate.
6. In tracing from long conjugate to source, pincushion distortion is desirable. This is because an element of area at the edge of the entrance pupil that is being illuminated should be a proportionately larger area of the source than one in the center. Unfortunately, most sources have less brightness at the edge than at the center.
7. Rarely are these systems made achromatic.

Figure 29-1 shows a fused quartz condenser at unit magnification. It fills a projection lens pupil of 0.5 inch. It is f/1 and will cover a film of

Table 29-1 Unit Magnification Condenser

	Radius	Thickness	Material	Diameter
0	0.00000	4.4990		0.500
1	0.00000	0.4815	Silica	2.600
2	−2.22208	0.0370	Air	2.600
3	2.22208	0.4815	Silica	2.600
4	0.00000	0.2470	Air	2.600
5	Aperture	3.8621	Air	2.000

Illumination Systems

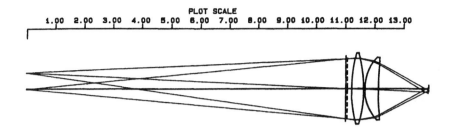

Figure 29-2 Fused quartz condenser 0.2×.

2 inch diameter. The distance from the object (lamp) to the first lens surface is 4.500.

The EFL is 2.424. The distance from the first lens surface to the image is 5.109. The program Design a Condenser in Appendix E is useful in designing such condensers. Although this condenser, shows considerable spherical aberration, it is a simple and effective condenser. Converting the spherical surfaces into parabolas will considerably improve it.

Figure 29-2 shows a quartz condenser at 0.2× magnification. It is also f/1 and will fill a 1 inch diameter projection lens entrance pupil.

Table 29-2 Fused Quartz Condenser 0.2×

	Radius	Thickness	Material	Diameter
0	0.00000	11.0330		1.000
1	Aperture	0.2000		2.000
2	3.63494	0.4212	Silica	2.180
3	−3.63494	0.0349	Air	2.180
4	1.31930	0.4541	Silica	1.920
5	6.26928	1.7556	Air	1.680

The EFL is 1.906. The distance from the object (projection lens' pupil) to the first lens surface is 11.233.

This has less spherical aberration than the previous system. Making surface 4 ($R=1.3193$) parabolic will reduce the spherical aberration.

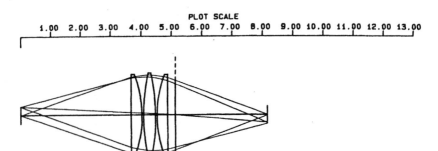

Figure 29-3 Pyrex condenser.

In Fig. 29-3 is shown an f/0.833 Pyrex condenser consisting of three spherical elements. It has unit magnification and will cover a film diameter of 1.8, which is adequate for a projector of 35 mm SLR film. The focal length is 2.059.

Table 29-3 Pyrex Condenser

	Radius	Thickness	Material	Diameter
0	0.00000	3.6877		0.500
1	0.00000	0.3500	Pyrex	2.360
2	–3.00000	0.0500	Air	2.360
3	5.00000	0.4000	Pyrex	2.450
4	–5.00000	0.0500	Air	2.450
5	3.00000	0.3500	Pyrex	2.360
6	0.00000	0.2500	Air	2.360
7	Aperture	3.0387	Air	2.000

The distance from the first lens surface to the image is 4.489. The distance from the object (lamp filament) to the first lens surface is 3.688. Note from the optical schematic that this last system is better corrected than the previous two. In both unit magnification systems we were tracing from the lamp to the projection lens pupil, whereas in the 0.2× magnification system, we trace from the projection lens to the lamp filament.

Illumination Systems 309

Frequently, the lamp filament (or arc) is placed at the center of curvature of a spherical mirror. This redirects energy that never would be collected by the condenser back to the source. However, the author's experience indicates that these mirrors only add a small amount to the illuminance. This is because in a modern tungsten halogen lamp the filament coils are very tightly packed together. This spherical mirror then images the filament onto itself, and little passes through.

Lamp #DYS is an example of a lamp with an open filament. It is a 600 watt lamp on a two-pin prefocus base. The mirror should be adjusted so that the filament image is a little above the lamp filament.

Some tungsten halogen lamps have these mirrors already built into the envelope. An example is ANSI code #BCK. This is a 500 watt lamp common in many slide projectors. It has a color temperature of 3250 K and a filament size of 0.425 × 0.405 inch.

Pyrex is a trade name of Corning Glass Works. A similar material is made by Schott (Duran) and by Ohara (E-6). It is the preferred choice for condenser lenses for use with lamps of moderate power. In order to remove excess infrared radiation, a heat-absorbing filter is often placed between the condensing lenses and the lamp. (Some designs even split the lens system and place this heat absorber inside the lens group.) Common heat filters are Schott KG series and Corning 1-58, 1-59, and 1-75. A recent trend is to use a dichroic heat reflector between the lamp and the lens system in place of the heat absorber. With modern coating technology, it has proven to be a more efficient system. These heat reflectors are available as stock items on 0.125 inch thick Pyrex material. They are used at normal incidence, or at 45 degree incidence (a cold mirror). This later case is generally more efficient, and it has the advantage of reflecting the heat outside of the optical system.

Figure 29-4 shows a condenser system designed to illuminate a transparency of 7.16 inch diameter. This transparency is then projected with a 9.54 inch focal length lens, f/8, at 10× Since this condenser system has a magnification of 2×, the source diameter is

$$\frac{9.54}{8(2)} = 0.596$$

The ANSI #BCK lamp may thus be used here.

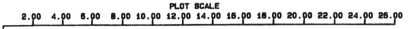
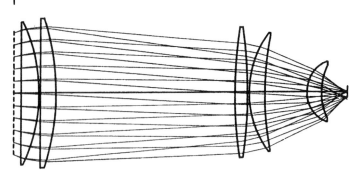

Figure 29-4 Condenser system for 10× projection.

Table 29-4 Condenser for 10× Projection

	Radius	Thickness	Material	Diameter
0	0.00000	1.10164E+01		1.192
1	0.00000	1.1760	BK7	8.400
2	−8.90300	0.1100	Air	8.400
3	0.00000	1.0490	BK7	8.780
4	−12.55440	11.9390	Air	8.780
5	23.46300	0.8390	Pyrex	7.700
6	−23.46300	0.1100	Air	7.700
7	5.40520	1.0300	Pyrex	7.010
8	15.92130	2.7680	Air	6.820
9	1.86540	0.9580	Pyrex	3.550
10	3.32290	1.7000	Air	3.160

The distance from the object (projection lens exit pupil) to the first lens surface is 11. The object-to-transparancy distance then is 10.5. The distance from the first lens surface to the image is 21.676. The focal length of the condenser system 110.028. The numerical aperture of this condenser is $7.16/10.5 = 0.6819$.

For heat resistance and economy, the rear elements are made of

Illumination Systems 311

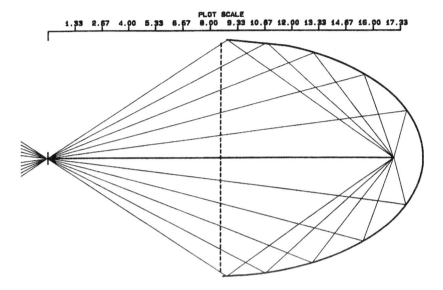

Figure 29-5 Xenon arc reflector.

Pyrex. However, Pyrex (as well as Duran and other similar materials) have considerable striae and some bubbles and inclusions. The front two elements, due to their proximity to the transparancy, are made from BK7. The large air space between these elements and the rear Pyrex lenses contains a folding cold mirror.

For very large xenon arc lamps (as in moving picture projection), a far more efficient system is the use of a deep ellipsoidal mirror. The lamp is placed along the mirror axis, with the arc being at one of the ellipsoid's foci. Such a system is capable of collecting far more energy than the above-described refractive devices. They have been made practical in recent years by the development of electroforming technology. In this process, a steel master is first made. It is then electroplated with nickel to perhaps a two mm thickness and then separated from the master. With proper controls, a low-stress part is produced. The inner concave surface then is aluminized.

Figure 29-5 shows an ellipsoidal reflector used with an arc source. Parameters for this reflector are:

Axial radius	2.7064
Ecentricity	0.853707
Vertex to arc	1.46
Vertex to image	18.50
Magnification	12.67

This should work well with a 7 K W xenon lamp. It has an arc size of 1 × 8 mm. The same concept of the source being placed with its longitudinal axis along the axis of an elliptical reflector is also used with small tungsten halogen lamps, ANSI #ENH and #BHB. These lamps have a dichroic coated reflector integral with the lamp and are extensively used in consumer audiovisual products.

REFERENCES

Brueggemann, H. P. (1968). Conic Mirrors, Focal Press, New York
Corning Glass Works (1984). Color filter glasses, Corning, NY 14831.
DuPree, D. G. (1975). Electroformed metal optics, *Proceedings of SPIE,* *65:*103.
General Electric (1988). Stage/studio lamps, SS- 123P, General Electric, Nela Park, Cleveland, OH 44112.
Hanovia (1988). Lamp data, Hanovia, 100 Chestnut St, Newark, NJ 07105.
Jackson, J. G. (1967). Light projection optical apparatus, U.S. Patent #3318184.
Koch, G. J. (1951). Illuminator for optical projectors, U.S. Patents #2552184 and #2552185.
Levin, R. E. (1968). Luminance, a tutorial paper, *SMPTE,77:*1005.
Optical Radiation (1988). Lamp data, Optical Radiation Corp., 1300 Optical Dr, Azusa, CA 91702.
Sharma, K. D. (1983). Design of slide projector condenser, *Applied Optics,* *22:*3925.
Smith, W. J. (1966). *Modern Optical Engineering,* McGraw-Hill, New York.
GTE Sylvania (1977). Sylvania lighting handbook, Danvers, MA 01923.
Wilkerson, J. (1973). Projection light source and optical system, U.S. Patent #3720460.

30
Lenses for Aerial Photography

Aerial photography is used for a wide variety of applications including military reconnaissance, aerial mapping, forest surveys, pollution monitoring, and various uses in oceanography. In order to record fine detail and to overcome the intrinsic grain size in film, long focal length lenses are combined with large format films. Standard widths of aerial films are 70 mm, 5 inch, and 9.5 inch. For dimensional stability an estar base film is preferred.

In order to reduce the effects of atmospheric haze, a filter is generally used. The purpose of this filter is to remove all short wavelength radiation to which most films are very sensitive. Ultraviolet and blue radiation is scattered more than green and red radiation by water vapor and dust particles in the atmosphere. Typically used filters are

    Wratten #2A,   opaque to wavelengths shorter than $0.41\mu m$
    3A,   opaque to wavelengths shorter than $0.44\mu m$
    4A,   opaque to wavelengths shorter than $0.46\mu m$

Solid glass filters are also used. For a 3 mm thickness the following apply:

    GG420,   opaque to wavelengths shorter than $0.41\mu m$
    GG455,   opaque to wavelengths shorter than $0.44\mu m$
    GG495,   opaque to wavelengths shorter than $0.47\mu m$

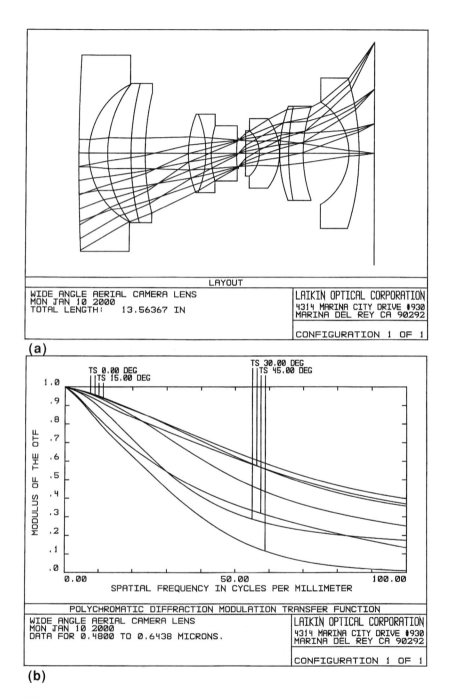

Figure 30-1 (a) Wide angle aerial camera lens. (b) Lens MTF. (c) Field curvature/distortion.

Lenses for Aerial Photography

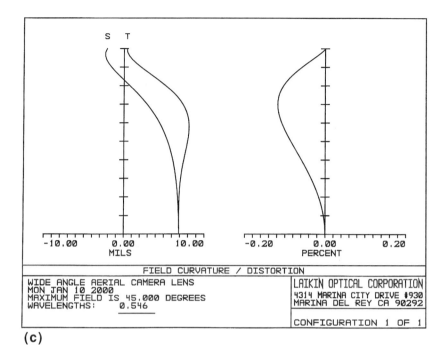

(c)

Figure 30-1 Continued

To be useful for photogrammetry, these lenses have less than 0.1% distortion.

Figure 30-1 shows a 5 inch focal length, f/4 wide angle aerial camera lens. It is a modification of Rieche (1991).

The distance from the first lens surface to the image is 13.564. The distortion curve for this lens reaches a maximum value of –0.12% at 0.7 of full field and has nearly zero distortion at full field. It is displayed in Fig. 30-1c.

Figure 30-2 shows a 12 inch focal length f/4 aerial camera lens that covers a 4 × 5 inch film format (6.4 diagonal).

The distance from the first lens surface to the image is 14.655. Distortion is negligible. FOV = 30 degrees. The reduced off-axis MTF response is due to intercept errors in the skew orientation.

Table 30-1 5-inch Aerial Camera Lens

Surf	Radius	Thickness	Material	Diameter
0	0.0000	1.000E+10		
1	92.3336	0.4604	FK5	8.800
2	3.4528	0.8787		6.160
3	5.1264	1.0619	LAF3	6.240
4	12.6041	0.6390	K5	6.240
5	11.8613	1.9753		5.420
6	4.6234	0.8039	LAKN13	3.520
7	−4.4757	0.2475	LLF1	3.520
8	4.3683	1.1785	LAKN13	2.480
9	7.5606	0.0881		1.320
10	Iris	0.3288		1.183
11	−64.8414	0.6324	FK51	2.040
12	−1.3538	0.9310	SF4	2.040
13	−2.5319	0.0125		3.120
14	5.9816	0.4039	SF4	4.220
15	10.7565	0.6655	LAK8	4.220
16	6.1021	2.0959		4.160
17	−2.6917	0.4604	SK5	4.600
18	−11.2615	0.7000		6.740
19	0.0000	0.0000		9.984

Table 30-2 12-inch Aerial Camera Lens

	Radius	Thickness	Material	Diameter
0	0.00000	1.000E+10		
1	6.44702	0.3864	SK16	4.350
2	17.24534	0.2034		4.200
3	4.96331	0.6852	SK16	3.880
4	17.73774	0.0702		3.500
5	48.46703	0.4095	F6	3.530
6	7.28391	0.0245		3.060
7	5.76310	0.4028	F6	3.060
8	3.42257	0.5841		2.620
9	Iris	3.6069		2.320
10	18.03505	0.4267	SK16	4.800
11	−67.91694	1.3394		4.800
12	16.08596	1.1250	SK16	5.450
13	−7.21560	0.2140		5.450
14	−8.33950	0.3612	SF1	5.280
15	−16.68175	2.4948		5.390
16	−6.72850	0.3607	F6	5.140
17	21.90122	1.9607		5.400

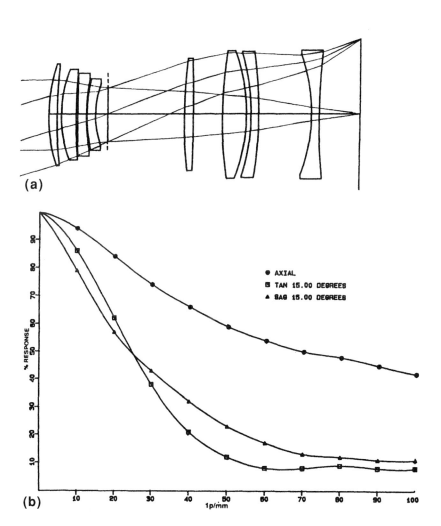

Figure 30-2 12-inch f/4 aerial camera lens. (b) Lens MTF.

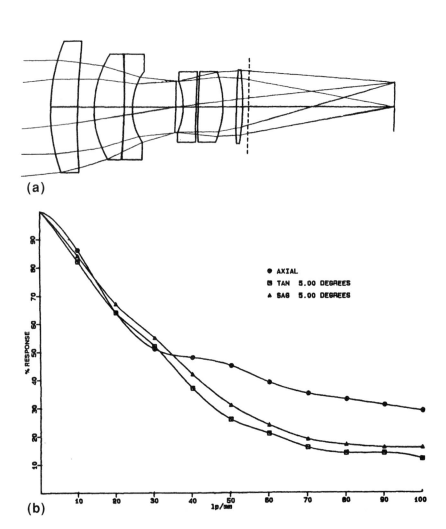

Figure 30-3 (a) 18-inch *f*/3 aerial lens. (b) Lens MTF.

Lenses for Aerial Photography 319

Some additional skew rays should have been traced in this skew orientation.

Figure 30-3 shows an 18 inch focal length f/3 aerial lens of 10 degree FOV. It covers a 2.25 inch square format (70 mm film).

Table 30-3 18-inch Aerial Camera Lens

	Radius	Thickness	Material	Diameter
0	0.00000	1.00E+10		
1	10.61548	1.9083	SK4	8.400
2	38.28208	1.1510		7.800
3	5.83702	2.1280	SK4	6.740
4	−42.26294	0.0139		6.740
5	−47.69489	0.5397	F5	5.830
6	3.75269	3.0564		4.580
7	Iris	0.5204		3.320
8	−4.27948	0.8918	LF5	3.380
9	32.19332	0.1553		3.940
10	31.29194	1.7507	SK4	4.530
11	−5.96022	0.9353		4.530
12	32.44910	0.4504	SK4	4.720
13	−16.52077	10.6248		4.720

The distance from the first lens surface to the image is 24.126. Distortion is 0.03%.

The small gap between the second and third lens surfaces and the fact that the radii are reasonably close to each other imply that perhaps this can become a cemented surface. This was tried, but with no success. This air surface has considerable third-order aberration contribution and so makes these elements sensitive to decentration and tilt errors.

Axial secondary color is considerable in this design. It causes the best image surface to move about 50 µm (away from the lens) from the image surface defined at 0.546 µm.

Figure 30-4 shows a 24 inch focal length f/6 lens for use with the

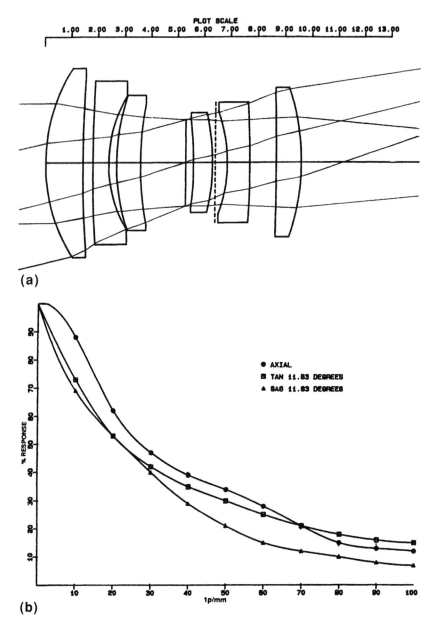

Figure 30-4 (a) 24-inch *f*/6 aerial lens. (b) Lens MTF.

Lenses for Aerial Photography 321

Table 30-4 24-inch Aerial Lens

	Radius	Thickness	Material	Diameter
0	0.00000	1.000E+10		
1	5.38182	1.3902	LAKN22	6.310
2	33.89543	0.3784		5.860
3	37.83808	0.6007	LF5	5.460
4	3.98992	0.2944		4.480
5	6.30323	0.9070	SK4	4.520
6	9.49052	1.7266		4.060
7	Iris	0.2957		2.880
8	−10.79902	0.7134	SK4	3.040
9	−6.66237	0.5534		3.330
10	−4.52682	0.8391	LF7	3.440
11	−29.80600	0.9761		4.020
12	−65.30257	0.9242	LAKN22	4.720
13	−6.96796	17.8836		5.000

9.5 inch wide film. The image size is 9×4.5 inches. This corresponds to a field of view of 23.66 degrees. Distortion is 0.076%.

The distance from the first lens surface to the image is 27.483.

Aerial cameras are quite large and expensive. They usually have the following features: interchangeable lenses, automatic exposure control, means of recording CRT data in a corner of the frame, and a focal plane shutter.

REFERENCES

Chicago Aerial Industries Data Sheets on Aerial Cameras, Chicago Aerial Industries, 550 W Northwest Highway, Barrington, IL 60010.

Hall, H. J. and Howell, H. K., ed. (1966). Photographic considerations for aerospace, Itek Corp., Lexington, MA.

Hoya Color Filter Glass Catalog, Hoya Optics, 3400 Edison Way, Fremont, CA 94538.

Kodak Data for Aerial Photography (1982). Pub. M-29, Eastman Kodak Co., Rochester, NY.

Kodak Aerial Films (1972). Pub. M-57, Eastman Kodak Co., Rochester, NY.

Kodak Wratten Filters, Pub. B-3 (1960). Eastman Kodak Co., Rochester, NY.
Properties of Kodak Materials for Aerial Photography (1974). Pubs. M-61, M-62, and M-63, Eastman Kodak Co., Rochester, NY.
Rieche, G. and Rische, G. (1991). Wide angle objective, U.S. Patent #5056901.
Schott Optical Glass Filters (1993). Schott Glass Technologies, Durea, PA.
Thomas, W., ed. (1973). *SPSE Handbook of Photographic Science and Engineering,* Wiley Interscience, New York.

31
Radiation-Resistant Lenses

Most optical glasses will darken (undergo browning) when exposed to x-ray or gamma radiation. This darkening is due to the presence of free electrons in the glass matrix. The darkening is not stable, in that it decreases with time. Raising the temperature and exposure to light accelerates this clearing process. Pure grades of fused silica (silicon dioxide) or quartz are very resistant to the effects of ionizing radiation.

Resistance to ionizing radiation may be increased in optical glasses by the addition of CeO_2 to the glass formula. This causes considerable absorption at 0.4 µm, but it substantially improves its radiation resistance. Most of the major optical glass manufacturers have cerium stabilized some of their most common optical glasses. Unfortunately, variety is limited and delivery sometimes slow.

The material from Schott have a "G" in their designation and a number identifying the CeO_2 concentration. The number represents 10×, in percent, the CeO_2 concentration. Thus BAK1G12 has 1.2% CeO_2.

Optical cements should not be used here. Most of the materials used for antireflection coatings are very resistant to radiation, so coatings should present no problem.

Figure 31-1 shows a 25 mm f/2.8 triplet made from radiation-resistant glasses. It was designed to be used with a one inch vidicon (0.625 inch diagonal).

The distance from the first lens surface to the image is 1.143. Distortion is 1.6%.

323

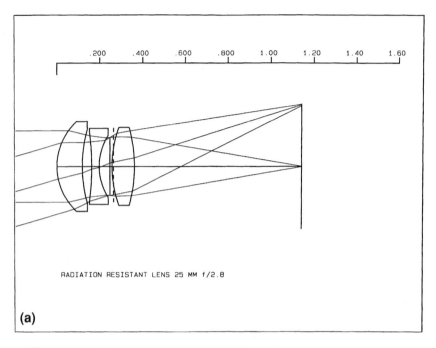

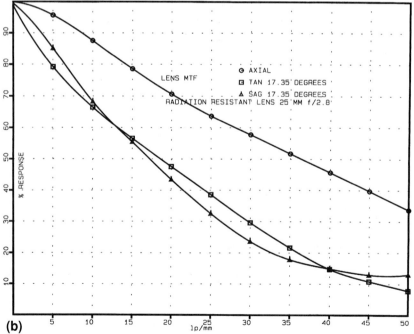

Figure 31-1 (a) 25 mm, f/2.8 radiation-resistant lens. (b) Lens MTF.

Table 31-1 Radiation Resistant Lens

	Radius	Thickness	Material	Diameter
0	0.00000	1.000E+07		
1	0.32506	0.1179	LAK9G15	0.450
2	0.74930	0.0438		0.370
3	−1.40604	0.0355	F2G12	0.350
4	0.30229	0.0510		0.310
5	Stop	0.0100		0.290
6	0.62453	0.1034	LAK9G15	0.390
7	−0.77895	0.7817		0.390

REFERENCES

Kircher, J. and Bowman, R., eds. (1964). *Effects of Radiation on Materials and Components,* Reinhold, New York.

Schott Glass (undated). Radiation resistant optical glass, Schott Glass Technologies, Duryea, PA.

32
Lenses for Microprojection

These optical systems are used by banks to record checks and other financial data, by credit card companies to record transactions, by libraries to make available many less used documents, by auto parts distributors to produce auto parts manuals conveniently, etc. The sensitive materials are either high-resolution diazo or conventional silver-based emulsions.

The projector has a light source and condensing system, a projection lens that images the film onto a high-gain screen, and a film transport mechanism. Some systems also have a pechan prism incorporated into the optical design to allow the user to rotate the image on the screen.

In Fig. 32-1 is shown a 24× microprojection lens. Track length (object-to-image distance) is 40. It is f/2.5 and so the NA = 0.192. It was designed to be used for a microfiche system. This format consists of 60 images on a film 5.826×4.134 (Williams, 1974), each image being 0.5×0.395 (0.637 diagonal).

At the top portion of the film is a title. An alternate version uses 98 images on the same size film.

The object distance to the first lens surface is 38.012. The distance from the first lens surface to the image is 1.987. Distortion = 0.7% (pincushion). The lens focal length = 1.543. Note that the next-to-last element, although of negative power, consists of a crown material. Ultrafiche is a system used with greater than 100× reduction (Fleischman, 1976b; Grey, 1970).

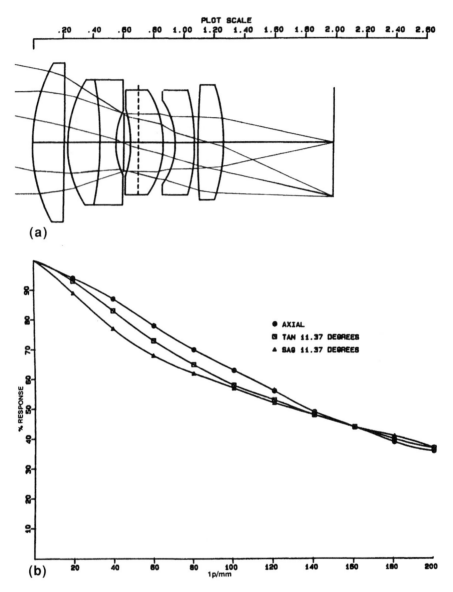

Figure 32-1 (a) 24× microprojection lens. (b) Lens MTF.

Lenses for Microprojection

Table 32-1 24× Microprojection Lens

	Radius	Thickness	Material	Diameter
0	0.00000	38.0116		
1	0.89262	0.2126	SSK4A	0.940
2	−49.08679	0.0209		0.940
3	0.62971	0.2137	SSKN5	0.740
4	−1.87244	0.1055	SF1	0.740
5	0.39596	0.0532		0.380
6	Stop	0.0438		0.340
7	−0.61216	0.2171	LAKN7	0.400
8	−0.51034	0.0800		0.620
9	−0.40081	0.1280	K10	0.500
10	−1.04802	0.0218		0.620
11	4.52618	0.1718	LAKN7	0.680
12	−0.98872	0.7183		0.680

REFERENCES

1. Fleischman, A. (1976a). Microfilm recorder lens, U.S. Patent #3998529.
2. Fleischman, A. (1976b). 7mm ultrafiche lens, U.S. Patent #3998528.
3. Grey, D. S. (1970). High magnification high resolution projection lens, U.S. Patent #3551031.
4. Olson, O. G. (1967). Microfilm equipment, *Applied Optics and Optical Engineering, 4:*167. Academic Press, New York.
5. SPSE (1968). Microphotography, Fundamentals and Applications, symposium held in Wakefield, MA, April 1968.
6. Williams, B. J. S. and Broadhurst, R. N. (1974). Use of microfiche for scientific and technical reports, AGARD Report #198.

33
First-Order Theory, Mechanically Compensated Zoom Lenses

In the discussions that follow, a zoom lens is an optical system in which the focal length (or magnification) varies, while the image plane remains stationary. For an excellent compendium of zoom lens papers, see Mann (1993). (In Chap. 38 we discuss variable focal length lenses in which the image plane substantially moves.) All of these systems are rotationally symmetric. It is desirable to divide the zoom region into logarithmically equal spaces: each succeeding focal length (or magnification) is obtained by multiplying the previous focal length by a constant. For example, for a 10× zoom lens, and four positions, this multiplier becomes the cube root of 10 (≈2.15443). In the data presented, this is only approximately true, since the desired value of the focal length (or magnification) was entered into the optimization program as a bounded item.

In Fig. 33-1 is shown a four-component mechanically compen-

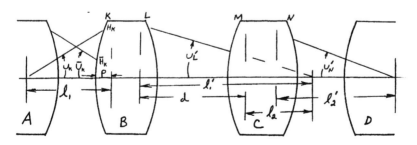

Figure 33-1 Mechanically compressed zoom lens.

sated zoom system. Lens group A is the front group. It only moves for focusing purposes. Thus the moving groups, B and C, always see a fixed virtual object for all conjugate distances. Group D is fixed. It generally contains the iris (the aperture stop) so that the image is always at a constant numerical aperture for all zoom positions.

Let us trace two paraxial rays from the object to the image. For convenience, we define one ray as emanating from an axial point and label the height it makes on a surface as H and its angle with the optical axis as U. The other ray will be a chief ray and has coordinates $\bar{H}$ and $\bar{U}$ (actually, any two paraxial rays will suffice). As shown in Fig. 33-1, $\bar{U}_K$ is a negative number, since the ray is falling.

The invariant, Ψ, is given by

$$\Psi = \bar{H}NU - HN\bar{U}$$

Since Ψ has the same value at each surface, we can evaluate it at the first lens surface. At this surface we make $\bar{H}$ zero, and so for a system in air,

$$\Psi = -H_1 \bar{U}_1$$

As a further simplification, we trace our paraxial rays so that $H_1 = 1$ and $\bar{U}_1 = 1$. Thus

$$\Psi = -1$$

The effective power of the B group is then given by

$$P_B = U_K \bar{U}'_L - \bar{U}_K U'_L$$

(Hopkins, 1962). Likewise for the C group P_C and the complete system P. As discussed above, the A group creates a virtual object for the B group. This distance from the surface K is

$$\frac{H_K}{U_K}$$

First-Order Theory, Mechanically Compensated Zoom Lenses 333

Likewise the image distance after going through group C from the surface N is

$$\frac{H_N}{U'_N}$$

The back focal length of the B group is found by

$$\text{BFL}_B = -\frac{[\overline{H}_L U_K - H_L \overline{U}_K]}{P_B}$$

The front focal length of the B group is found by

$$\text{FFL}_B = \frac{[\overline{H}_K U'_L - H_K \overline{U}'_L]}{P_B}$$

And, of course, the same is true for the C group. The distance P that the principal plane is from surface K is

$$\frac{1}{P_B} + FFL_B$$

(FFL_B is negative if the image is to the left). So $l_1 = -[H_K/U_K + 1/P_B + FFL_B]$ since l_1 is a negative number as shown. The magnification of the B and C groups is $M = U_K/U'_N$. d, the separation between the B and C groups (measured between principal planes) is given by

$$d = \frac{P_B + P_C - Pb_C}{P_B P_C} = T(L) + \frac{1}{P_B} - \text{BFL}_B - \frac{1}{P_C} + \text{FFL}_C$$

Let S be the distance between object and image (minus the principal plane separations of the B and C groups:

$$S = d + l'_2 + l_1$$

This is to remain constant as we zoom. Following the nomenclature of Clark (1973), we denote the original positions by l and the new position by L. We now move the B and C lens groups to new positions. Their separation is D.

$$L_1' = \frac{L_1}{1+P_B L_1} \tag{1}$$

$$L_2' = \frac{L_2}{1+P_C L_2} \tag{2}$$

$$M = \frac{L_1' L_2'}{L_1 L_2} = \frac{1}{(1+P_B L_1)(1+P_C L_2)} \tag{3}$$

$$S = -L_1 + L_1' - L_2 + L_2' \tag{4}$$

Substituting Eqs. (1) and (2) into Eq. (4) we obtain

$$S = \frac{-P_B L_1^2}{1+P_B L_1} - \frac{P_C L_2^2}{1+P_C L_2} \tag{5}$$

From Eq. (3) we get obtain

$$\frac{1}{1+P_C L_2} = M(1+P_B L_1)$$

Substituting into Eq. (5) we obtain

$$L_2 = \frac{1 - M(1+P_B L_1)}{M P_C (1+P_B L_1)} \tag{6}$$

Substituting Eq. (6) into Eq. (5) we obtain,

$$S(1+P_B L_1) = \frac{-P_B L_1^2 - [1 - M(1+P_B L_1)]^2}{M P_C}$$

Solving for L_1 we obtain

$$L_1 = \frac{-b \pm (b^2 - 4ac)^{1/2}}{2a} \tag{7}$$

where $a = MP_BP_C + M^2P_B^2$, $b = SMP_BP_C - 2MP_B + 2M^2P_B$, and $c = SMP_C + M^2 - 2M + 1$. The sign for the square root term in Eq. (7) is so chosen that it would yield the starting solution values $L_1 = l_1$. The value for the sign then is

$$\frac{2al_1 + b}{\sqrt{b^2 - 4ac}}$$

which takes on a value of ±1. From Eq. (1) we find a new value of L'_1. From Eq. (5) we find a new value of L_2. The separation between the B and C groups then becomes

$$D = L'_1 - L_2$$

Again D is measured between principal planes of the B and C groups. Since these principal planes do not change locations with respect to their groups as the lens assembly zooms, the new zoom spacers become

$$\text{TH1} = T(K-1) + l_1 - L_1$$
$$\text{TH2} = T(L) + D - d$$
$$\text{TH3} = T(K-1) + T(L) + T(N) - \text{TH1} - \text{TH2}$$

where TH1 is the zoom space between the A and B groups. Likewise, TH2 corresponds to the B–C space, and TH3 is for the C–D space.

Clark (1973) gives a different and thin lens analysis for this mechanically compensated system. The program Zoom Spacings in Appendix E computes these spacings per the above equations.

REFERENCES

Chunken, T. (1992). Design of a zoom system by the varifocal differential equation, *Applied Optics, 31*:2265.

Clark, A. D. (1973). *Zoom Lenses,* American Elsevier, New York.
Grey, D. S. (1973). Zoom lens design, *Proc. SPIE, 39:*223.
Hanau, R. and Hopkins, R. E. (1962). Optical design, MIL-HDBK-141, Sec. 6, Standardization Division, Defense Supply Agency, Washington DC.
Kingslake, R. (1960). The development of the zoom lens, *J. SMPTE, 69:*534.
Yamaji, K. (1967). Progress in Optics, Vol. 6, Chap. 4, Design of zoom lenses, Academic Press, New York.

34
First-Order Theory, Optically Compensated Zoom Lenses

In an optically compensated zoom lens, all of the moving lens groups move in unison, as if they were linked together on a common shaft. This is shown in Fig. 34-1. It has the distinct advantage that no cam is required. However, such systems are generally longer than a comparable mechanically compensated system. Most important, the image plane moves slightly with zoom. An object of the designer is to make this movement less than the depth of focus.

In Fig. 34-1, lens groups A, C, and E are fixed, while groups B and D are linked together. Tracing a paraxial, axial ray, it makes an angle with the optical axis after refraction at surface M of U'_M and

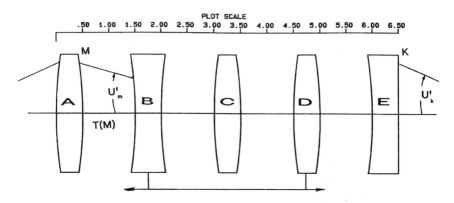

Figure 34-1 Optically compensated zoom lens.

emerges at the last surface K with an angle U'_K. Tracing a paraxial chief ray through the system is denoted by $\overline{U}'_M$, etc. The axial space between surfaces M and $M + 1$ is $T(M)$.

$$\frac{dU'_K}{dT(M)} = \frac{NU'_M[\overline{U}'_K U'_M - U'_K \overline{U}'_M]}{\Psi}$$

(Hanau and Hopkins, 1962). Ψ is the invarient $= N\overline{H}U - NH\overline{U} = -H_0\overline{U}$ for an object in air. The zoom ratio

$$Z = \frac{[U'_K]_1}{[U'_K]_2}$$

where the 1 denotes the system at its initial position and the 2 denotes the system at its final zoom position. Since we are dealing with derivative changes, the total change in U'_K is the sum of these changes for all the variable air spaces.

As discussed above, the image moves with zoom. Wooters and Silvertooth (1965) have shown that the maximum number of positions where the longitudinal image motion is zero is equal to the number of moving air spaces. One must realize that for a distant object, if the first group moves, this is not considered a variable air space. Also this is the maximum number of crossing points; an actual system may very well have fewer. As an extreme example, we may consider a zoom lens with only one moving element. Figure 34-2 shows such a thin lens arrangement for a distant object. Paraxially, we obtain

$$\text{EFL} = \frac{F_A F_B}{F_A + F_B - D} \quad \text{BFL} = \frac{(F_A - D)F_B}{F_A + F_B - D}$$

For the image to be stationary, $D + \text{BFL} = $ a constant. If we move the B lens to a new position we obtain new values for EFL, BFL, and of course D. Let the new EFL be noted as $Z(\text{EFL})$ where Z is the zoom ratio. For the sum $D + \text{BFL}$ to be the same as previously,

First-Order Theory, Optically Compensated Zoom Lenses

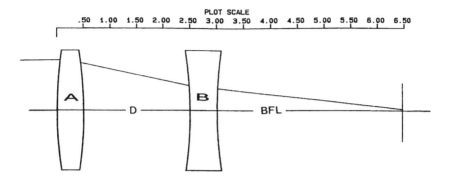

Figure 34-2 Zoom with one moving lens.

$$Z = \frac{F_A^2}{\text{EFL}^2}$$

The new value of D would be

$$D_2 = F_A + F_B - \frac{F_A F_B}{Z(\text{EFL})}$$

The image would be in the same location at these two positions. Johnson and Feng (1992) discuss an afocal infrared lens system with one moving group. They call it mechanically compensated. By this definition, it is optically compensated, since the image remains exactly stationary at only two positions. However, if the cams are so arranged as to move both lens groups in relation to the image plane, then obviously the image will remain stationary. See Chap. 39 for variable focal length lenses.

REFERENCES

Back, F. G. and Lowen, H. (1954). The basic theory of varifocal lenses with linear movement and optical compensation, *JOSA, 44*:684.

Bergstern, L. (1958). General theory of optically compensated varifocal systems, *JOSA, 48:*154.

Hanau, R. and Hopkuns, R. E. (1962). MIL-HDBK-141, Optical design, Defense Supply Agency, Washington, DC 6-26.

Jamieson, T. H. (1970). Thin lens theory of zoom systems, *Optica Acta, 17:*565.

Johnson, R. B. and Feng, C. (1992). Mechanically compensated zoom lenses with a single moving element, *Applied Optics, 31:*2274.

Mann, A. (1993). Selected papers on zoom lenses, SPIE MS85, SPIE, Bellingham, WA.

Pegis, R. J. and Peck, W. G. (1962). First-order design theory for linearly compensated zoom systems, *JOSA, 52:*905.

Wooters, G. and Silvertooth, E. W. (1965). Optically compensated zoom lens, *JOSA, 55:*347.

35
Mechanically Compensated Lenses

In Fig. 35-1 is shown a 10× mechanically compensated zoom lens covering the range 0.59 to 5.9 (15 to 150 mm). It is f/2.4 and has a 0.625 (16 mm) diameter image. The lens prescription (at the short focal length) is given in Table 35-1. It is designed for an object at infinity for a 1 inch vidicon.

This lens is of the form +, −, +, +. That is, all the groups are positive except for the first moving group, surfaces 6 through 10. The variable air spaces are given in the following table. The distance from the first lens surface to the image is 12.222.

In Fig. 35-1a the lens is shown in its short focal length position. Figure 35-1b shows the movement of the lens groups. $T(5)$ and $T(10)$ are large monotonic functions. $T(15)$ is nearly linear but only has a small movement.

MTF data for this lens is displayed in Figs. 35-1c through 35-1f. These correspond to the four focal positions of Table 35-2. If this lens is used for 16 mm cinematography, the field of view will be slightly reduced. For example,

16 mm format has a 12.7 mm diagonal
Super 16 format has a 14.5 mm diagonal

If used with a 1 inch vidicon (which has a 16 mm diagonal), then one should add the camera tube face plate glass to the lens prescription. This is generally a low index of refraction crown glass (1.487) (see Appendix

341

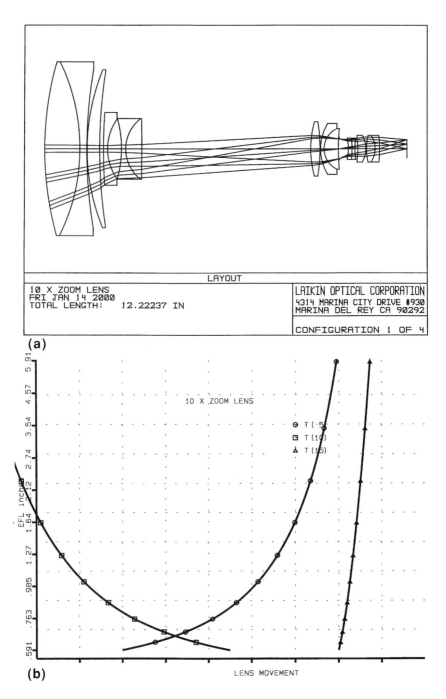

Figure 35-1 (a) 10× zoom lens. (b) Lens MTF. (c-f) Polychromatic diffraction MTF.

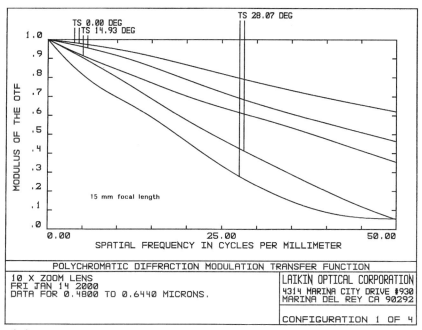

(c)

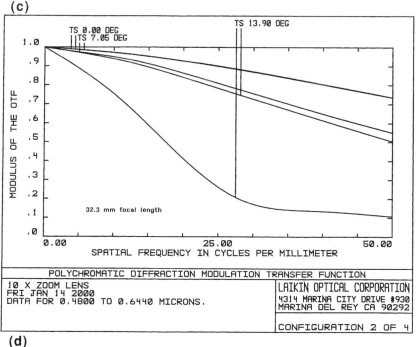

(d)

Figure 35-1 Continued

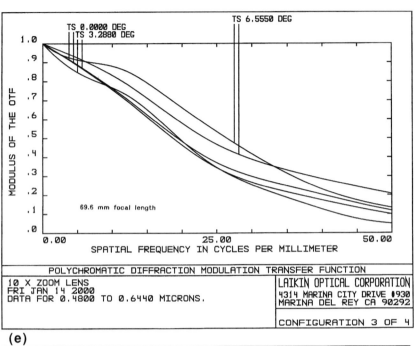

(e)

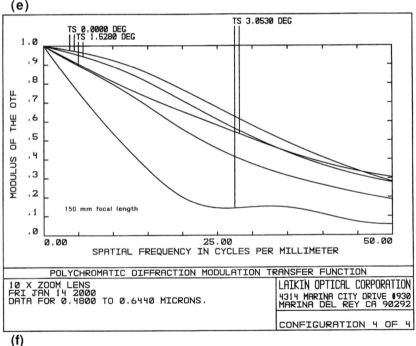

(f)

Figure 35-1 Continued

Table 35-1 0.59 to 5.9 f/2.4 Zoom Lens

Surf	Radius	Thickness	Material	Diameter
0	0.0000	1.000E+10		
1	10.1098	1.1937	LAK9	5.720
2	−6.3839	0.2360	SF5	5.720
3	15.4899	0.0100		5.260
4	6.6703	0.4254	LAFN21	5.200
5	16.0779	0.1335		5.100
6	17.6988	0.1400	LAK9	2.400
7	1.6653	0.4664		1.960
8	−3.7115	0.1400	LAK9	1.960
9	1.2520	0.5434	SF5	2.000
10	−50.0672	5.6883		2.000
11	3.3233	0.3154	LAK9	1.800
12	−5.3099	0.0195		1.800
13	1.6006	0.1169	SF1	1.680
14	0.7352	0.4803	LAK9	1.340
15	6.8782	0.0432		1.340
16	Iris	0.2947		0.639
17	−1.2146	0.1191	LAK8	0.620
18	−3.9731	0.0984	SF1	0.700
19	1.9568	0.0955		0.660
20	−2.4821	0.2734	SF8	0.680
21	−0.8604	0.0243		0.880
22	1.9237	0.2929	BAK4	0.880
23	−0.7665	0.1550	SF1	0.880
24	−3.5517	0.9170		0.880
25	0.0000	0.0000		0.598

Table 35-2 Focal Length vs. Air Spaces

EFL	T(5)	T(10)	T(15)	Pupil	Distortion (%)
.591	0.133	5.688	.043	−3.763	−7.17
1.272	2.435	3.234	.195	−9.434	0.71
2.741	4.013	1.431	.421	−17.060	2.51
5.905	5.095	.010	.760	−29.596	3.28

A). Most color cameras now use a prism beam-splitter to separate the red, green, and blue images (Cook, 1973). Reoptimization in this case would be required. The program ZOOMCAM in Appendix E is useful in fabricating the cams for these mechanically compensated zoom lenses.

Many low-power microscopes, used in inspection, as surgical microscopes, in electronic assembly, etc., use an afocal pod as a means of obtaining the zoom. This has the advantage of forming modular components. That is, the objective, eyepiece, and prism assembly are the same as on the fixed power microscopes. The afocal pod is inserted to zoom. Table 35-3 gives the prescription for such an afocal system. It has a zoom ratio of four.

The system is shown in Fig. 35-2a in the $M=2$ power position. The distance from the first to the last surface is 2.935.

Table 35-3 Afocal Zoom for a Microscope

Surf	Radius	Thickness	Material	Diameter
0	0.0000	1.0000E+10		
1	Stop	−2.3622		0.400
2	1.2388	0.1504	K10	0.760
3	9.3542	0.0234		0.760
4	1.3062	0.0787	SF1	0.700
5	0.6690	0.1145	PSK3	0.700
6	2.4729	0.6310		0.600
7	2.6233	0.0709	SK16	0.460
8	0.2835	0.1794	SSKN5	0.460
9	0.9740	0.3496		0.360
10	−0.6347	0.0606	SK16	0.320
11	3.7396	0.3475		0.320
12	−1.7308	0.0492	SK16	0.400
13	0.6212	0.1614	SF1	0.500
14	2.3140	0.1352		0.440
15	38.9045	0.0789	SF1	0.880
16	1.5054	0.1923	PSK3	0.880
17	−2.5537	0.0362		0.880
18	5.2888	0.2755	BK7	0.880
19	−1.7936	0.0000		0.880

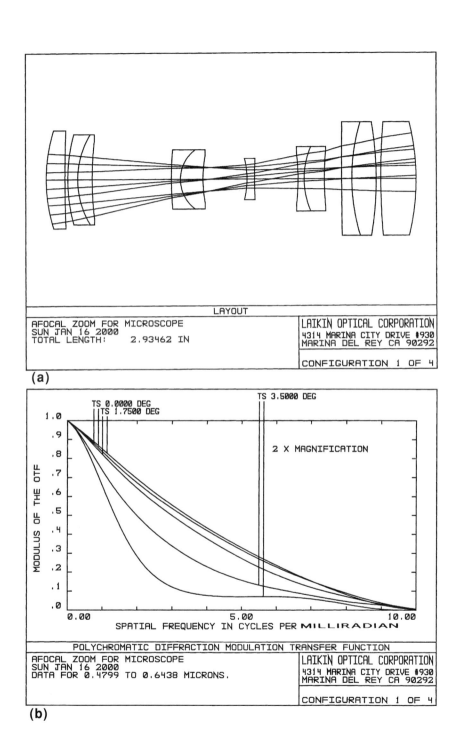

Figure 35-2 (a) Afocal zoom for microscope. (b-f) Lens MTF.

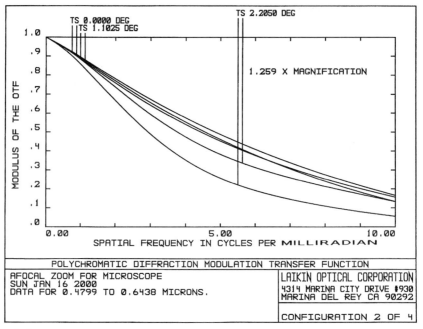

(c)

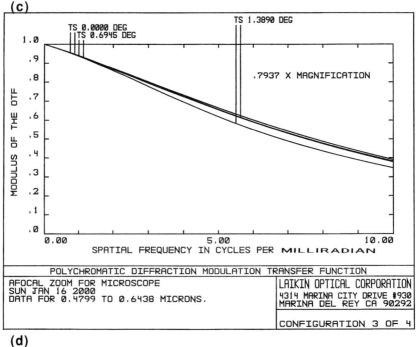

(d)

Figure 35-2 Continued

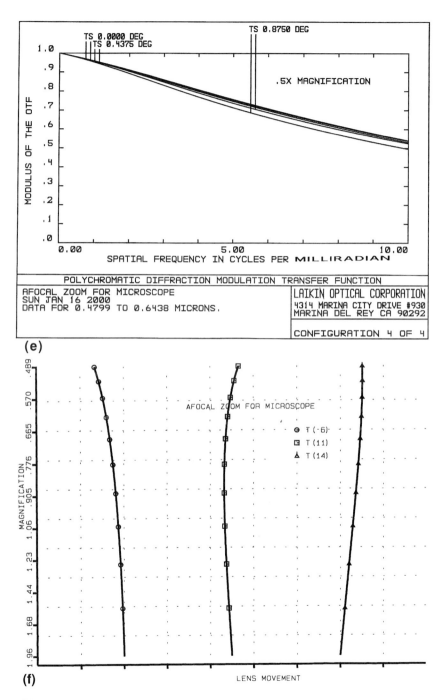

Figure 35-2 Continued

Table 35-4 Afocal Zoom Movements

M	θ	T(6)	T(11)	T(14)	Distortion (%)
2	3.500	0.631	0.347	0.135	0.86
1.259	2.205	0.497	0.186	0.431	0.01
0.7937	1.389	0.255	0.237	0.622	−0.07
0.5	0.875	0.008	0.453	0.652	−0.03

Note: A paraxial trace of the above system yields an angular magnification of 1.955.

The computer program used, as well as most programs, employs a "perfect" paraxial surface after the last real lens surface. This has the effect of placing the infinitely located image at its paraxial location. Image errors now may be considered in the normal manner. In this case the paraxial surface had a focal length of 1000 mm. Thus 1p/mm in the MTF analysis becomes cycles/milliradian. This MTF data is presented in Figs. 35-2b through 35-2e. Table 35-4 gives the zoom movements at the various magnifications; θ is the semi field angle, in degrees of the entering beam.

The above zoom movements are plotted in Fig. 35-2f. From this we note that $T(14)$ moves in a nearly linear fashion, while $T(11)$ reverses direction at about 0.78 magnification.

In Fig. 35-3a is shown an infrared Cassegrain zoom lens. It has an intermediate image inside the primary mirror. It was designed to cover the wavelength region 3.2 to 4.2 μm. The relative weightings are as in Table 35-5.

Table 35-5 Relative Weight vs. Wavelength

Wavelength (μ)	Weight
3.2	0.3
3.4	0.6
3.63	1.0
3.89	0.6
4.2	0.3

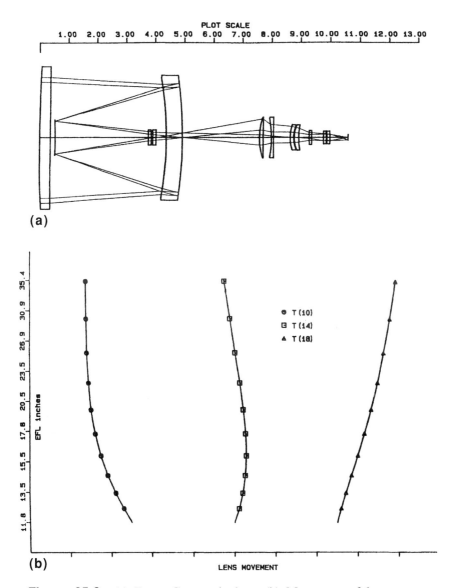

Figure 35-3 (a) Zoom Cassegrain lens. (b) Movement of lens groups. (c) 300 mm focal length. (d) 432.6 mm focal length. (e) 624 mm focal length. (f) 899.9 mm focal length.

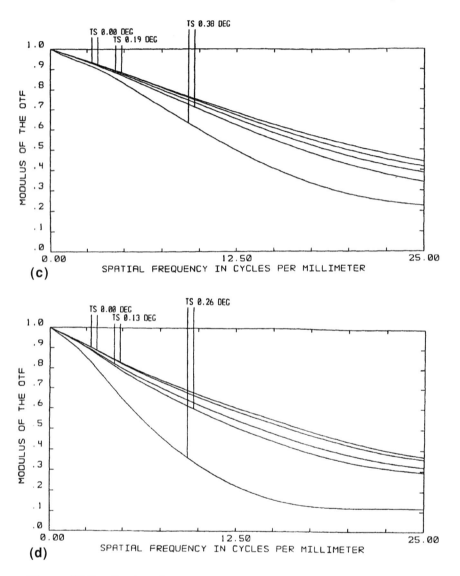

Figure 35-3 Continued

Mechanically Compensated Lenses 353

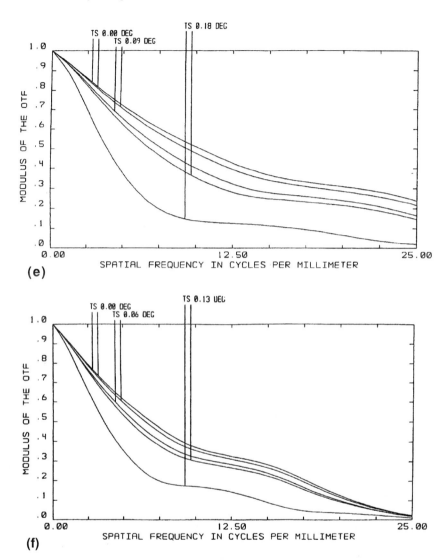

Figure 35-3 Continued

The entrance pupil has a constant 3.937 diameter and is located 22.462 to the right of the first lens surface. This places the aperture stop in contact with the secondary mirror. This has the effect of minimizing the diameter required (and hence the obscuration) of the secondary mirror. The image diameter is 0.157 (4 mm).

IRTRAN2 is a sintered and hot pressed form of ZnS. ZnSe made

Table 35-6 Infrared Zoom Cassegrain

Surf	Radius	Thickness	Material	Diameter
0	0.0000	0.100000E+11		0.00
1	41.7742	0.3797	IRTRAN2	4.330
2	0.0000	3.9956		4.330
3	−9.9443	0.5500	IRTRAN2	3.760
4	−12.1154	0.0000	Mirror/IRTRAN2	3.860
5	−9.9443	−3.8239		3.700
6	Stop	3.2207		1.100
7	−36.2852	0.1000	Germanium	0.450
8	−2.3898	0.0553		0.450
9	−2.7359	0.1000	Silicon	0.450
10	2.5240	3.5389		0.450
11	1.7824	0.1245	Silicon	1.260
12	16.4118	0.2891		1.260
13	−2.6837	0.0756	Germanium	0.940
14	0.0000	0.5701		1.260
15	1.0989	0.1181	Germanium	0.780
16	0.8329	0.0207		0.650
17	0.8891	0.1502	Silicon	0.780
18	1.7684	0.3616		0.780
19	0.0000	0.0788	Silicon	0.390
20	1.1734	0.4048		0.390
21	3.2380	0.1000	Silicon	0.390
22	−1.3825	0.0264		0.390
23	−0.9210	0.1000	Germanium	0.390
24	−1.1487	0.6101		0.390
25	0.0000	0.0000		0.180

Table 35-7 Zoom Movements

Focal length	T(10)	T(14)	T(18)	Distortion (%)
11.81	3.539	0.570	0.362	−3.5
17.03	2.872	0.836	0.762	−0.2
24.57	2.467	0.709	1.295	1.1
35.43	2.406	0.279	1.786	2.7

by a chemical vapor deposition process can also be used (with only a minor adjustment to the prescription), since its index and dispersion are reasonably close. The distance from the front lens vertex to the image is 10.596.

The movement of these lens groups is plotted in Fig 3b. MTF data at the focal lengths indicated in Table 35-7 is plotted in Figs. 35-3c through 35-3f.

Rifle sights and binoculars are afocal devices. However, they differ from the afocal pods used in microscopy in that they are not components but rather complete systems. For use as a rifle sight, a system must have very long eye relief to allow for the recoil of the rifle. Also long length is desirable because such length allows easy mounting on the rifle.

In Fig. 35-4a such a zoom rifle scope is shown in the low magnification position. The power will change from 2.349 to 7.375. The entrance pupil is in contact with the first lens surface. The exit pupil is a constant diameter of 0.197 (5 mm). This is more than adequate to accomodate most outdoor viewing scenes. Focusing for the individual shooter is accomplished by moving the eyepiece, and the front objective is moved to focus at various object distances. Eye relief is 3.543 (90 mm). The zoom ratio of 3.14 is a little too small. It should be increased. This system suffers from considerable astigmatism, particularly in the low magnification position.

The distance from the first to the last surface is 12.551. The entrance pupil is in contact with the first lens surface. The objective images the distant target onto the first surface of the reticle (surface 7). This is often a tapered cross line and is adjusted to the bore sight of the

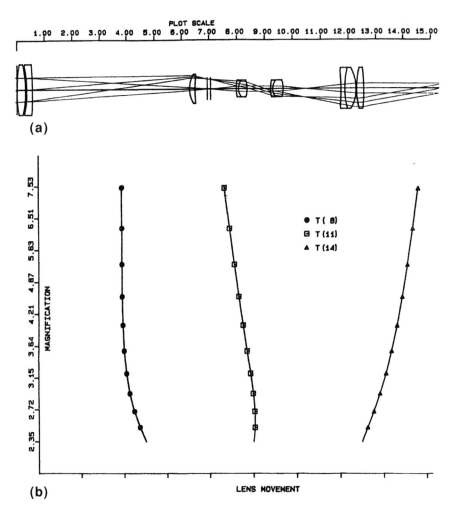

Figure 35-4 (a) Zoom rifle scope lens. (b) Zoom lens movements. (c) Zoom rifle scopes m=2.349. (d) Zoom rifle scope m = 3.556. (e) Zoom rifle scope m=5.143. (f) Zoom rifle scope m=7.375.

Mechanically Compensated Lenses

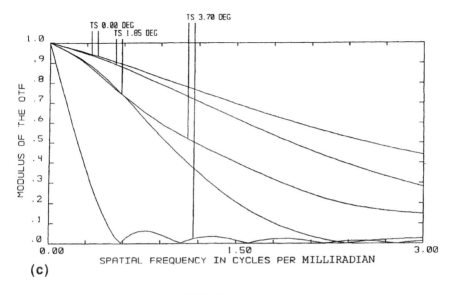

(c)

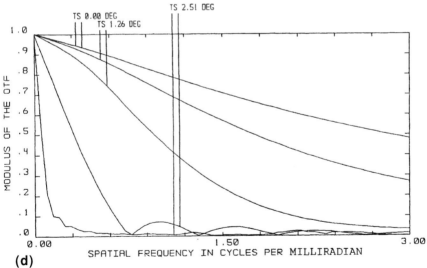

(d)

Figure 35-4 Continued

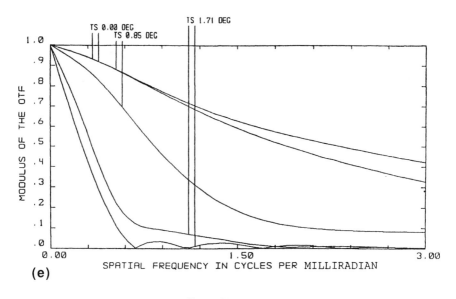

(e)

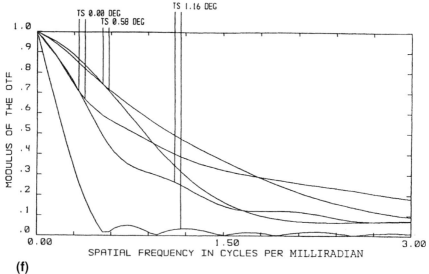

(f)

Figure 35-4 Continued

Mechanically Compensated Lenses 359

Table 35-8 Zoom Rifle Scope

Surf	Radius	Thickness	Material	Diameter
0	0.0000	1.0000E+10		
1	4.0394	0.3138	BAK2	1.654
2	−3.5518	0.0521		1.654
3	−3.3885	0.2002	SF1	1.575
4	−10.3062	5.7204		1.654
5	0.8467	0.1969	K5	0.945
6	3.2325	0.4504		0.945
7	0.0000	0.1181	K5	0.709
8	0.0000	0.9254		0.709
9	1.2332	0.0787	SF4	0.551
10	0.3915	0.3145	SSK2	0.551
11	−0.8397	0.8349		0.551
12	1.1365	0.0787	SF4	0.551
13	0.3899	0.3754	SSK2	0.551
14	−0.8309	2.0650		0.551
15	−5.5677	0.1378	SF4	1.102
16	2.2165	0.4087	BK7	1.339
17	−1.3068	0.0197		1.339
18	2.5341	0.2601	SK15	1.339
19	−5.4323	3.5433		1.339

Table 35-9 Zoom Movements for Rifle Scope

Mag	θ	$T(8)$	$T(11)$	$T(14)$	Distortion (%)
2.349	3.70	0.9254	0.8349	2.0650	4.8
3.556	2.51	0.4988	0.7823	2.5442	1.7
5.143	1.71	0.3655	0.5138	2.9460	0.9
7.375	1.16	0.3545	0.1621	3.3086	1.2

rifle. Since the zoom action occurs after the reticle, any misalignment as a result of the moving lenses will not affect the aiming accuracy.

θ is the full field semifield angle in degrees. Mag is the angular magnification. These zoom movements are plotted in Fig. 35-4b. MTF data at the four zoom positions are given in Figs. 35-4c through 35-4f. Note that this data applies to the emerging ray bundle: what the eye sees. Thus 3 cycles/milliradian corresponds to ray angles of 1/3000 = 1 minute of arc. We then note that resolution on axis is more than adequate at all zoom positions. At the full field angle, resolution is seriously degraded for the sagittal component of the MTF.

In Fig. 35-5a is shown (in the high magnification position) a system suitable as a stereo microscope. It has a zoom range of three, effective focal length varying from 0.591 to 1.77 (15 mm to 45 mm). Table 35-10 gives the prescription when in the long focal length (low magnification) position.

Table 35-10 Stereo Microscope

Surf	Radius	Thickness	Material	Diameter
0	0.0000	1.0000E+10		
1	1.8446	0.1017	SF1	0.331
2	0.8936	0.1220	SK14	0.331
3	−1.5598	0.0197		0.331
4	1.5598	0.1220	SK14	0.331
5	−0.8936	0.1017	SF1	0.331
6	−1.8446	5.5073		0.331
7	0.0000	0.9843	BK7	0.394
8	0.0000	0.6890		0.394
9	Stop	1.0210		0.394
10	1.0373	0.1451	SF1	0.575
11	0.6255	0.2101	FK5	0.575
12	−3.4421	1.5684		0.575
13	−0.8369	0.0748	K7	0.472
14	0.3053	0.1649	SF4	0.472
15	0.4539	0.4174		0.472
16	4.7394	0.1682	FK5	0.906
17	5.9833	0.0197		0.906
18	1.3819	0.0906	SF4	0.906
19	0.7705	0.2719	K7	0.906
20	−1.6815	7.1763		0.906

Mechanically Compensated Lenses

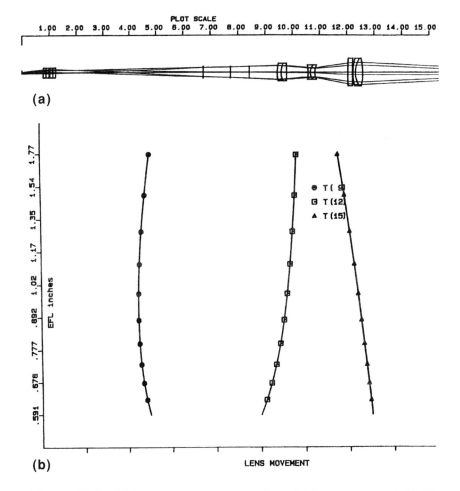

Figure 35-5 (a) Stereo zoom microscope lens. (b) Lens movement. (c) 15 mm focal length. (d) 22.15 focal length. (e) 31.2 mm focal length. (f) 45 mm focal length. (g) 60° deviation prism.

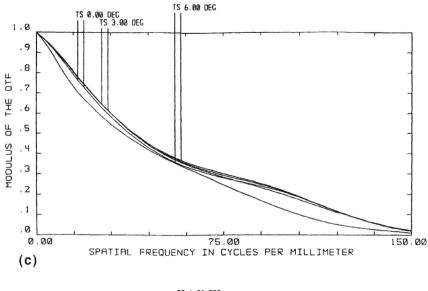

(c)

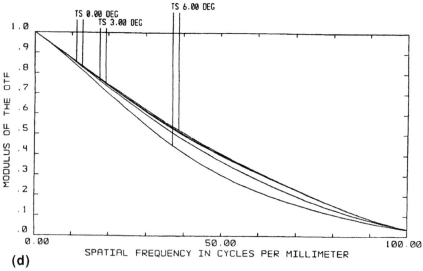

(d)

Figure 35-5 Continued

Mechanically Compensated Lenses

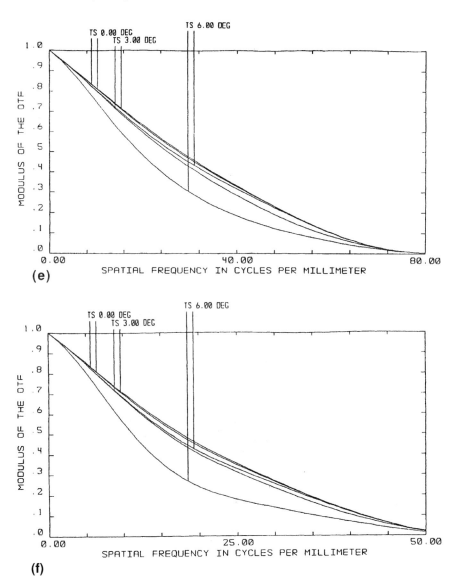

Figure 35-5 Continued

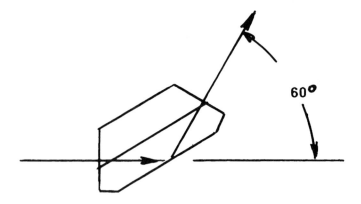

(g)   60 DEVIATION PRISM

Figure 35-5 Continued

The distance from the eye (the entrance pupil) to the first lens surface is 0.787. These first lenses (surfaces 1–6) are a Plossl eyepiece with a focal length of 0.828. It forms an intermediate image 0.676 from surface 6. A glass block of BK7 (surfaces 7 and 8) is actually a roof prism (Hopkins, 1962), which inverts and reverts the image. It also deviates the beam so that the eyepiece assembly is 60 degrees from the vertical axis. This makes for convenient viewing. A 60 degree deviation prism is shown in Fig. 35-5g. Surface 9 is the aperture. It is fixed in location and so is part of the A group. This means that the eye position remains fixed as we zoom. Surfaces 10–12 are the B moving group and 13–15 are the C group. The distance from the first lens surface to the image is 18.976. Table 35-11 provides data for the various zoom positions.

Magnification for the eyepeice portion is given by

$$Mag = \frac{EFL + 10}{EFL} = 10.83$$

Table 35-11 Zoom Positions for Stereo Microscope

EFL	T9	T(12)	T(15)	Distortion (%)
0.591	1.016	0.789	1.202	−0.8
0.872	0.800	1.155	1.052	−0.7
1.228	0.776	1.416	0.814	−0.7
1.772	1.021	1.568	0.417	−1.2

For the complete system, the magnification varies from 6.6 to 17.9. The zoom movements are plotted in Fig. 35-5b. In Figs. 35-5c through 35-5f are given MTF data at the various zoom settings.

Determining the axial resolution from these graphs and then dividing by the focal length gives a nearly constant angular resolution, as seen by the eye, of 0.0005 radians over the field. This corresponds to about 1.6 minutes of arc.

In Fig. 35-6a is shown a zoom microscope with magnification changing from 11 to 51. It is shown in the low magnification position. Lens prescription is listed in Table 35-12.

Table 35-13 lists effective focal lengths, entrance pupil data, and lens group spacings.

Notice that surface 13 is the aperture stop of the system. It has a fixed diameter over the zoom region. Since it is part of a moving group, the entrance pupil moves slightly with zoom and should not be objectionable. Surface 6 is a field stop. It has a fixed diameter of 0.420. It yields a fixed field of view for all zoom positions of 20 degrees.

MTF data is shown in Figs. 35-6c through 35-6f.

Note that both of these zoom microscope systems use separate front objectives. This has the advantage of allowing each arm to be inclined to the work. Another method (Murty, 1997) uses a large front objective common to both sides. This has two advantages: simpler construction and easy changing of the overall magnification of the zoom system. This is shown in Fig. 35-6g.

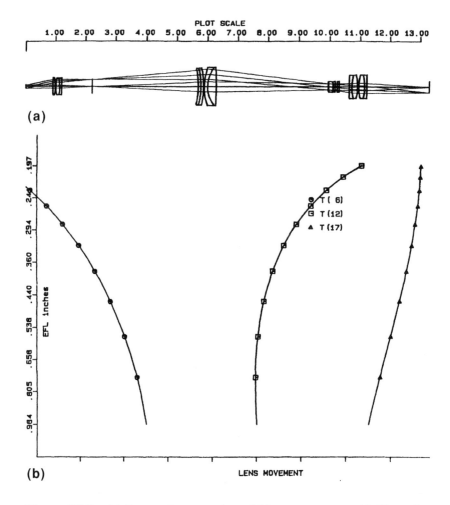

Figure 35-6 (a) Zoom microscope lens. (b) Lens movement. (c) 25 mm focal length. (d) 14.6 mm focal length. (e) 8.56 mm focal length. (f) 5.0 mm focal length. (g) Stereo microscope with common front objective.

Mechanically Compensated Lenses

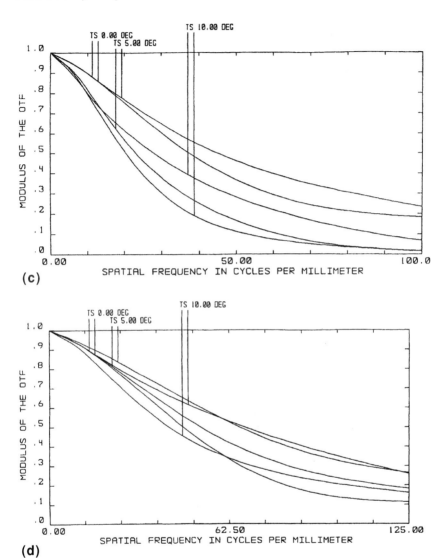

Figure 35-6 Continued

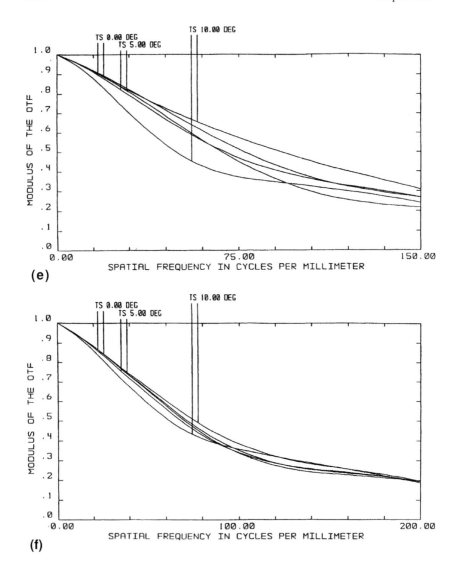

Figure 35-6 Continued

Mechanically Compensated Lenses

Stereo microscope with common front objective

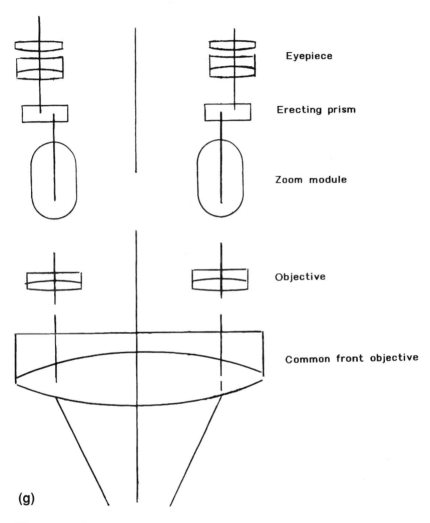

Figure 35-6 Continued

Table 35-12 200m Microscope 11-51×

Surf	Radius	Thickness	Material	Diameter
0	0.0000	1.0000E+10		
1	5.1837	0.0834	SK5	0.504
2	−1.0536	0.0198		0.504
3	1.0605	0.1310	PK2	0.504
4	−0.9616	0.0591	SF1	0.504
5	4.4730	1.0221		0.504
6	0.0000	3.4759		0.420
7	−1.9918	0.0866	SF1	0.992
8	−1.8161	0.0866	BAK1	1.102
9	−2.1348	0.0197		1.102
10	1.5302	0.1049	SF1	1.142
11	0.8165	0.2953	BAK1	1.142
12	−7.3527	3.6710		1.142
13	−1.0239	0.1530	SF1	0.176
14	−0.5613	0.0598	BAK1	0.201
15	−4.4802	0.0672		0.315
16	−3.5666	0.0600	BAK1	0.315
17	5.2239	0.3385		0.315
18	21.6742	0.0787	F5	0.701
19	1.1738	0.1969	FK5	0.701
20	−1.1066	0.0315		0.701
21	1.1066	0.1969	FK5	0.701
22	−1.1738	0.0787	F5	0.701
23	−21.6742	2.0732		0.701

The distance from the first lens surface to the image is 12.390.

Table 35-13 Lens Spacings for 10-51× Zoom

EFL	Entrance diameter	Pupil distance	$T(6)$	$T(12)$	$T(17)$
0.984	0.098	0.906	3.476	3.671	0.338
0.576	0.082	0.911	2.528	3.867	1.091
0.337	0.060	0.857	1.323	4.738	1.424
0.197	0.037	0.872	0.031	5.953	1.502

These lens movements are plotted in Fig. 35-6b.

Mechanically Compensated Lenses

In Fig. 35-7a is a zoom lens suitable for use with a vidicon, CCD, or 16 mm film camera (16 mm diameter image). It is f/2 and varies in focal length from 0.787 to 4.33 (20 to 110 mm). This design is a modification of the Kato (1989) patent. Note that for the usual video camera usage, a glass block should be added to the rear

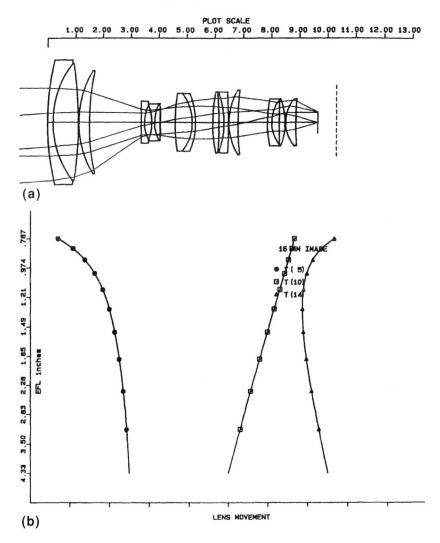

Figure 35-7 (a) 20 to 110 mm zoom lens. (b) Lens movement. (c) 20 mm focal length. (d) 35.3 mm focal length. (e) 62.3 mm focal length. (f) 110 mm focal length.

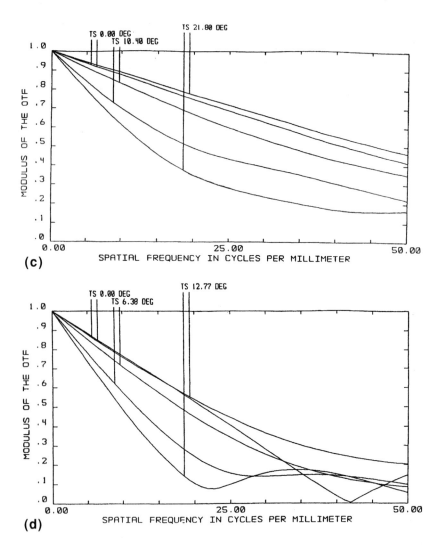

Figure 35-7 Continued

Mechanically Compensated Lenses 373

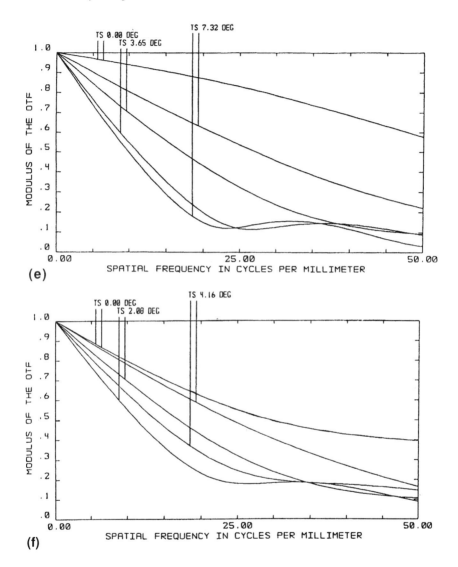

Figure 35-7 Continued

of the lens system (as in the patent). Note that unlike the above zoom camera lens designs, the iris is located between the two moving groups. Moving the iris into the forward portion of the lens tends to reduce the front lens diameters. This is particularly important at the wide field position. The iris also moves, which causes an increased complexity in the mechanical construction. Figure 35-7a

Table 35-14 20-110mm Zoom

Surf	Radius	Thickness	Material	Diameter
0	0.0000	0.100000E+11		0.00
1	7.8280	0.1969	SF6	3.992
2	2.8307	0.9140	LAK9	3.740
3	−10.7459	0.0197		3.740
4	2.9002	0.3832	LAK9	3.307
5	6.9126	1.8613		3.228
6	0.0000	0.1181	LAFN28	1.339
7	1.1254	0.2587		1.102
8	−1.3799	0.0984	LAK9	1.102
9	1.1861	0.1903	SF57	1.181
10	0.0000	0.0175		1.181
11	Iris	0.5199		0.93
12	5.1084	0.6153	BK7	1.843
13	−1.3631	0.1181	SF8	1.843
14	−2.1947	0.6104		1.843
15	4.2413	0.2257	LAF2	1.953
16	−11.3549	0.1752		1.953
17	−2.5532	0.1575	SF57	1.85
18	−16.1840	0.0197		1.953
19	1.8858	0.3128	LAK9	2.063
20	8.2502	1.0418		2.063
21	3.1532	0.1575	SF6	1.496
22	1.0632	0.2655		1.417
23	18.9826	0.2100	BK7	1.496
24	−1.8327	0.0197		1.496
25	1.0689	0.3605	LAK10	1.496
26	7.1817	0.7870		1.496

The distance from the first lens surface to the image is 9.655.

shows the lens in its long focal length position. Additional clearance between the iris and surface 10 should be added. Table 35-14 lists the lens prescription (long focal length setting).

Movement of the lens groups is plotted in Fig. 35-7b. Note that the movement of the C group (surfaces 12–14) reverses direction. The following table gives the entrance pupil locations, group positions, and distortion data for the various focal lengths.

Table 35-15 Zoom Lens Movements

EFL	Entrance diameter	Pupil distance	$T(6)$	$T(12)$	$T(17)$	Distortion (%)
0.787	−2.465	0.050	2.035	0.146	0.778	−3.1
1.390	−5.631	1.046	0.909	0.988	0.066	6.5
2.453	−8.418	1.575	0.295	1.100	0.039	7.0
4.331	−10.315	1.861	0.018	0.520	0.610	5.6

MTF data is plotted in Figs. 35-7c through 35-7f.

In Fig. 35-8a is shown a 25-to-125 mm focal length zoom lens for 35 mm motion picture (Academy format, 1.069 diagonal) cinematography. It is f/4. This system differs from the previous examples in that it has three moving groups. Although only two moving groups are required to change focal length and maintain a fixed image plane, the additional moving group is used for aberration control. This design is a modification of the Cook (1972) patent. As discussed in this patent, focusing is accomplished by moving the doublet and singlet following the front two singlets. The overall length of the lens then does not change with focus as it does in most zoom lenses. Another advantage with this system is that a nearly constant angular field of view is obtained as one focuses from distant to near objects. Table 35-16 lists the lens prescription for a distant object at the short focal length setting. To focus on near objects, T4 is decreased.

Table 35-17 lists zoom movements and distortion vs focal length.

Table 35-16 25-125mm Zoom

Surf	Radius	Thickness	Material	Diameter
0	0.0000	0.100000E+11		0.00
1	13.0839	0.4481	LAK21	5.100
2	7.5719	0.0861		4.560
3	8.8048	0.3177	LAK21	4.620
4	4.7224	0.8616		4.160
5	−46.3801	0.7096	SF6	4.040
6	−5.8725	0.5362	LF5	4.040
7	11.3528	0.6574		3.400
8	−5.0519	0.2708	LAK10	3.440
9	−10.7181	0.5442		3.700
10	0.0000	0.2676	SF6	3.940
11	9.3521	0.6051	SK16	3.940
12	−6.8485	0.0197		3.940
13	9.3534	0.2676	SF6	4.040
14	4.9965	0.5994	LAK21	4.040
15	−16.9145	0.0197		4.040
16	3.7464	0.5004	LAK21	3.900
17	17.1248	0.1071		3.800
18	6.2694	0.1417	LAFN21	2.060
19	1.5955	1.0782		1.740
20	−2.4228	0.1169	LAF3	1.620
21	2.0118	0.7414	SF6	1.900
22	−16.3788	3.2611		1.900
23	30.9694	0.1260	SF6	1.960
24	2.3413	0.4365	SSKN5	1.960
25	−2.8669	0.0197		1.960
26	2.8661	0.3012	FK5	1.920
27	−5.8595	0.0269		1.920
28	Iris	0.1017		0.966
29	−2.1793	0.0792	LAFN23	0.980
30	1.4875	0.1887	SF6	1.100
31	25.3790	0.5406		1.100
32	−7.5399	0.3504	PK2	1.140
33	−2.2715	0.4459		1.260
34	10.9873	0.0906	SF6	1.280
35	1.3382	0.2751	PSK2	1.280
36	−3.3161	3.0844		1.280

The distance from the front surface to the image is 18.224.

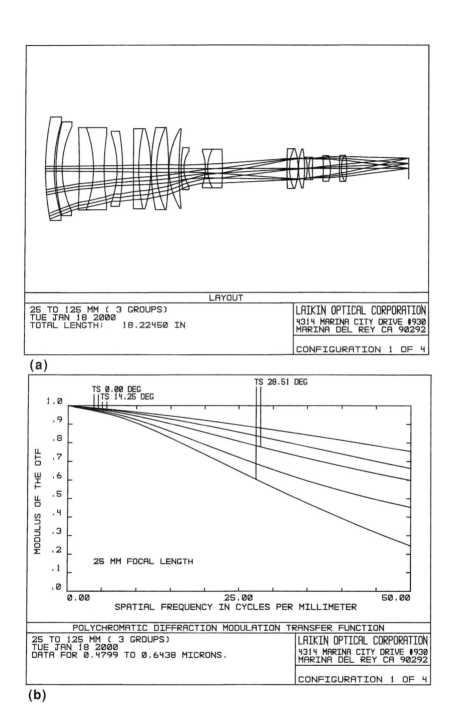

Figure 35-8 (a) 25 to 125 mm focal length zoom lens for 35mm cinematography. (b-e) Lens MTF.

377

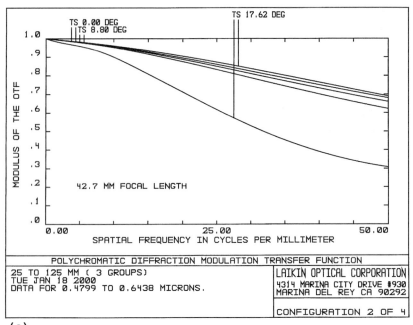

(c)

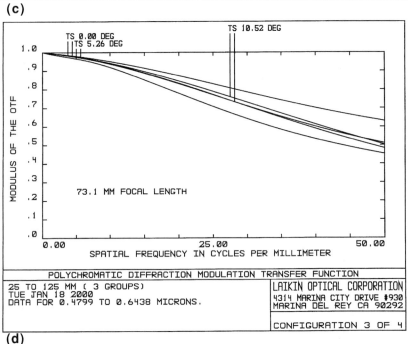

(d)

Figure 35-8 Continued

Mechanically Compensated Lenses 379

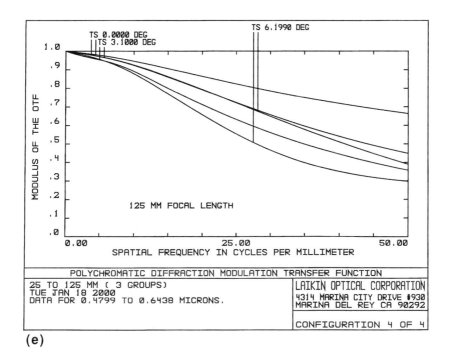

(e)

Figure 35-8 Continued

Table 35-17 Lens Movements for 25-125mm Zoom

EFL	Entrance diameter	Pupil distance	$T(6)$	$T(12)$	$T(17)$	Distortion (%)
0.984	−4.660	0.544	0.107	3.261	0.027	−5.12
1.683	−5.699	0.621	1.050	1.920	0.348	−0.04
2.878	−7.127	0.515	1.829	0.860	0.735	1.24
4.921	−9.865	0.010	2.601	0.039	1.289	1.53

MTF data is plotted in Figs. 35-8b thru 35-8e.

In Fig. 35-9a is shown a lens suitable for TV usage. The focal length can be made to vary from 12 to 234 mm. It was designed for use with a 1/2 inch CCD (6.4 × 4.8 mm). Following is the lens prescription in the short focal length configuration.

Table 35-18 12-234mm Zoom

Surf	Radius	Thickness	Material	Diameter
0	0.0000	1.0000E+10		
1	6.7937	0.3000	SF56A	4.900
2	3.9082	0.8330	PSK53A	4.620
3	−87.2079	0.0100		4.620
4	3.8585	0.7975	FK51	4.400
5	8.9733	0.0047		4.100
6	4.3545	0.0900	F2	2.120
7	1.3078	0.2174		1.760
8	3.0849	0.4000	SF57	1.820
9	−2.4100	0.0900	LAF2	1.820
10	2.7985	0.2156		1.420
11	−2.3266	0.0900	LAK8	1.420
12	10.1497	3.0904		1.380
13	−1.0960	0.0630	SSKN8	0.760
14	0.8686	0.1034	SF6	0.860
15	2.1800	0.7088		0.860
16	Iris	0.1181		1.021
17	6.5649	0.1689	LASFN30	1.240
18	−1.9870	0.0100		1.240
19	8.3231	0.1059	LAK9	1.240
20	−6.4489	0.0090		1.240
21	2.1218	0.2534	FK5	1.240
22	−1.4584	0.0700	SF6	1.240
23	9.4126	0.3978		1.140
24	1.2410	0.4000	FK51	1.240
25	12.7718	0.2673		1.040
26	0.9389	0.0945	LAFN7	0.940
27	0.6195	0.3105		0.800
28	1.1200	0.3000	PSK3	0.840
29	−4.1366	0.5000		0.840
30	0.0000	0.1378	BK7	0.388
31	0.0000	0.0000		0.336

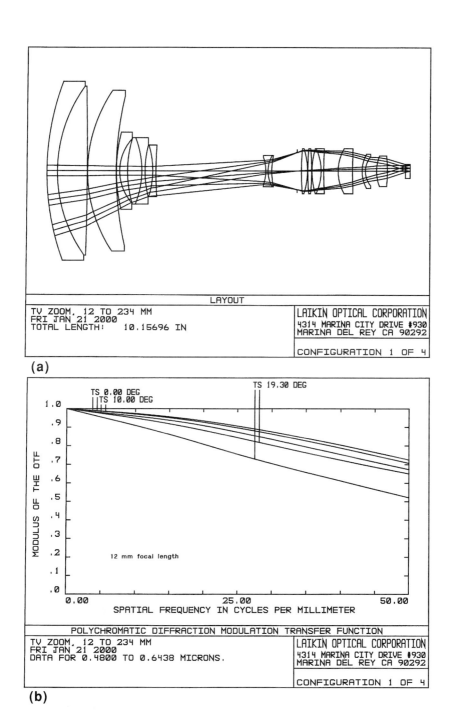

Figure 35-9 (a) 12 to 234 mm television zoom lens. (b-f) Lens MTF.

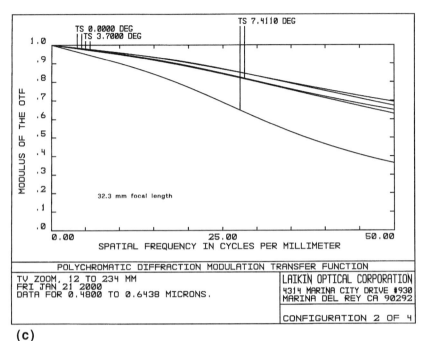

(c)

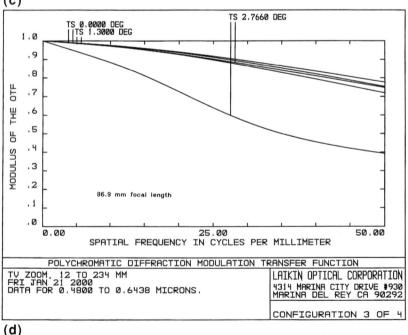

(d)

Figure 35-9 Continued

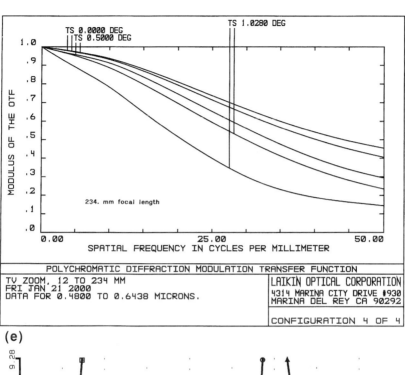

(e)

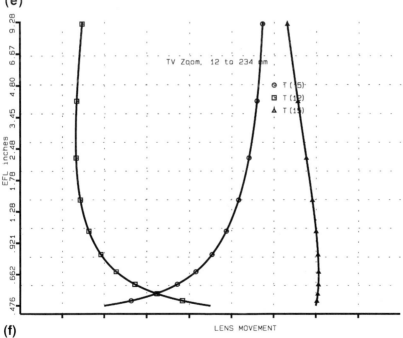

(f)

Figure 35-9 Continued

Table 35-19 Lens Movements 12-234mm Zoom

EFL	T(5)	T(12)	T(15)	f#	Distortion (%)
0.4724	0.0047	3.0904	0.7088	1.6	2.01
1.2716	1.8745	1.1581	0.7714	1.6	0.13
3.4227	3.0763	0.0703	0.6574	2.0	0.34
9.2126	3.7261	0.0585	0.0194	2.8	0.97

The distance from the first surface to the image is 10.157. Note that the f# increases as one zooms from short to long focal length. This was done to reduce the size of the front group. The BK7 plate at the rear of the lens is part of the CCD array. To yield even illumination at the CCD array, this lens is nearly telecentric. (Exit pupil is located − 10.36 from the image.) It is a modification of the Enomoto (1998) patent. MTF data is plotted in 35-9b–e. Zoom movements are plotted in figure 35-9f.

REFERENCES

Betensky, E. (1992). Zoom lens principles and types, *Lens Design,* SPIE Critical Review, Vol. CR31:88.

Cook, G. H. (1959). Television zoom lenses, *J. SMPTE, 68:*25.

Cook, G. H. and Laurent, F. R. (1972). Objectives of variable focal length, U.S. Patent #3682534.

Cook, G. H. (1973). Recent developments in television optics, *Royal Television Society J.*, 158.

Enomoto, T. and Ito, T. (1998). Zoom lens having a high zoom ratio, U.S. Patent #5815322.

Hopkins, R. E. (1962). Optical design, MIL-HDBK-141, pp. 13–47, Standardization Division, Defense Supply Agency, Washington, D.C.

Jamerson, T. H. (1971). Zoom lenses for the 8–13 micron region, *Optica Acta, 18:*17.

Johnson, R. B. (1990). All reflective four element zoom telescope, International Lens Design Conference, Monterey, CA, p. 669.

Kato, M., Tsuji, S., Sugiura, M., and Tanaka, K. (1989). Compact zoom lens, U.S. Patent #4854681.

Kojima, T. (1970). Zoom objective lens system with highly reduced secondary chromatic aberration, U.S. Patent #3547523.

Macher, K. (1970). High speed varifocal objective system, U.S. Patent #3549235.

Macher, K. (1974). High speed varifocal objective, U.S. Patent #3827786.

Mann, A. (1992). Infrared zoom lenses in the 1980s and beyond, *Optical Engineering, 31:*1064.

Murty, A. S. et al. (1997). Design of a high resolution stereo zoom microscope, *Optical Engineering 36:*201.

Nothnagle, P.E. and Resenberger, H. D. (1969). Zoom lens system for microscopy, U.S. Patent #3421807.

Rah, S. Y. and Lee, S. S. (1989). Spherical mirror zoom telescope satisfying the aplanatic condition, *Optical Engineering, 28:*1014.

Schuma, R. F. (1962). Variable magnification optical system, U.S. Patent #3057259.

Yahagi, S. (2000). Zoom lens for a digital camera, U.S. Patent #6014268.

36
Optically Compensated Zoom Lenses

In Fig. 36-1a is shown an optically compensated zoom lens with two moving groups. It zooms from 3.94 to 7.87 (100 to 200 mm) focal length and covers the full-frame 35 mm motion picture format (1.225 diagonal) It is shown in the long focal length position.

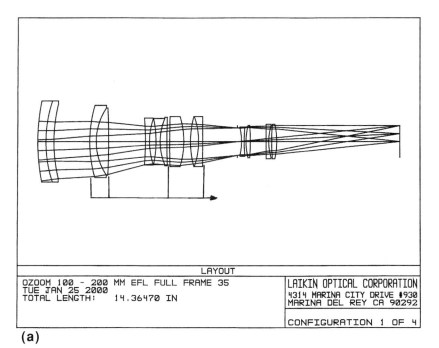

(a)

Figure 36-1 (a) Optically compensated zoom lens. (b–e) Lens MTF. Figure continues on pages 389 through 391.

Table 36-1 Lens Prescription 100 to 200 mm Focal Length, Optical Zoom

Surf	Radius	Thickness	Material	Diameter
0	0.0000	1.0000E+10		
1	7.8491	0.3238	LF5	3.140
2	4.7883	0.3238	LAK8	3.140
3	5.8979	1.4358		2.900
4	4.9300	0.2568	SF4	2.800
5	2.2971	0.4961	BAFN10	2.600
6	−24.6447	1.4590		2.600
7	−11.1125	0.1634	SK14	1.820
8	2.7388	0.2578		1.740
9	−4.1362	0.1633	SK14	1.740
10	2.2929	0.2655	SF4	1.795
11	29.7026	0.1111		1.880
12	47.4998	0.5507	SK14	1.940
13	−3.2744	0.2670		1.940
14	3.6694	0.3179	SK14	1.940
15	−2.8881	0.1619	SF4	1.940
16	0.0000	1.3775		1.940
17	Iris	0.1680		1.072
18	−2.2685	0.1053	BAF4	1.080
19	5.9067	0.0729		1.140
20	2.5231	0.1694	BASF2	1.260
21	−8.6344	0.6323		1.260
22	125.1504	0.1053	SF2	1.340
23	2.1238	0.1147		1.240
24	6.9089	0.2034	LAK8	1.340
25	−2.9516	4.8619		1.340

The distance from the first lens surface to the image is 14.365.

Table 36-1 gives the lens prescription at this long focal length.

Table 36-2 gives the lens positions and distortion at four focal positions.

Figures 36-1b through 36-1e show the MTF for the above four focal settings. Note that although the paraxial BFL does vary with focal length, this MTF data is plotted on a common image surface as indicated in Table 36-1.

Optically Compensated Zoom Lenses

Table 36-2 Lens Positions and Distortion vs. Focal Length

EFL	T(3)	T(6)	T(11)	T(16)	BFL	Distortion(%)
3.937	2.710	0.021	1.549	0.104	4.861	−0.52
4.960	2.291	0.518	1.052	0.523	4.860	0.19
6.250	1.855	0.992	0.578	0.959	4.858	0.67
7.874	1.436	1.459	0.111	1.378	4.858	0.99

The system as shown has no vignetting and is f/5 throughout the zoom. Distortion is fairly small.

In Fig. 36-2a is shown an optically compensated zoom lens (shown in its long focal length position) suitable for use with a single lens reflex (SLR) camera (43.3 mm diagonal). The focal length varies from 2.835 (72 mm) to 5.709 (145 mm). It is f/4.5 through the zoom

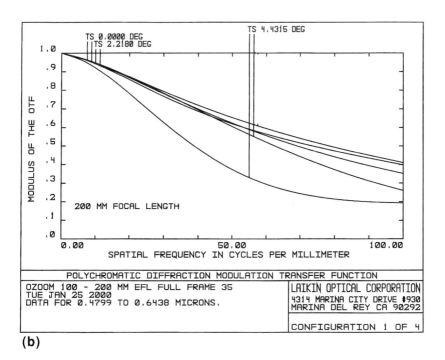

(b)

Figure 36-1 Continued

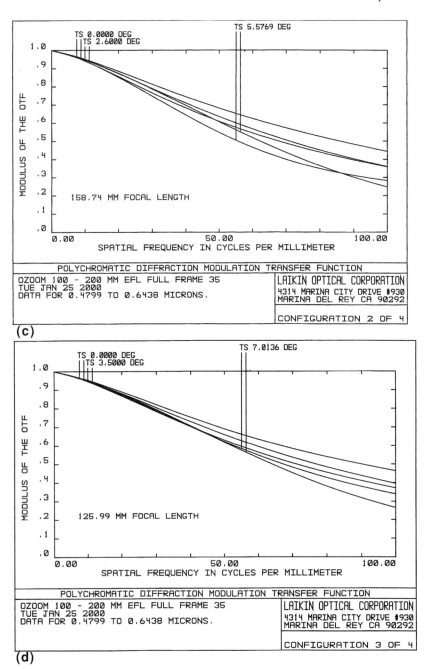

Figure 36-1 Continued

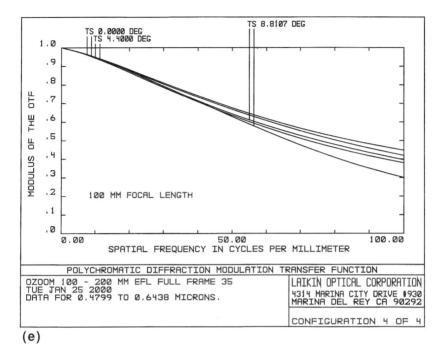

(e)

Figure 36-1 Continued

range. Following is the lens prescription in the short focal length position. It is a modification of Ikemori (1980) (Table 36-3).

Table 36-4 provides the spacings, distortion, and entrance pupil locations at the various focal lengths. Note that the lens changes its length as it zooms.

MTF data is shown in Figs. 36-2b through 36-2e

The lens movements and variation in back focal length are plotted in Fig. 36-2f.

Note that additional clearance should be allowed for the iris. Also, as a consumer product for a single lens reflex camera, most photographers would feel that this lens is too long. Mechanical compensation, although having the complexity of a cam, does provide a substantially more compact lens system.

In Figure 36-3a is shown an optically compensated lens used to project an image onto an 8 inch diameter screen. The distance from the

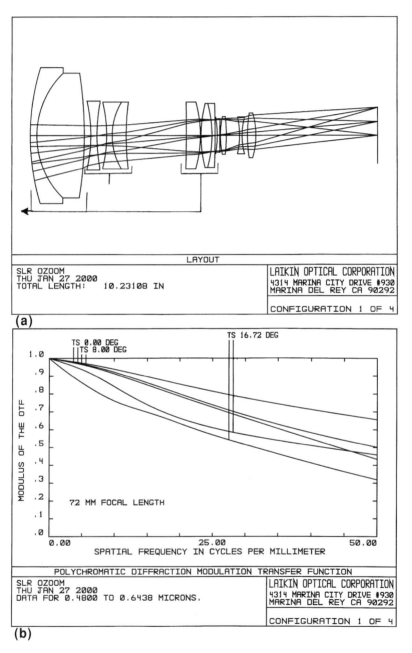

Figure 36-2 (a) Optically compensated zoom lens for SLR camera. (b–f) Lens MTF.

Optically Compensated Zoom Lenses

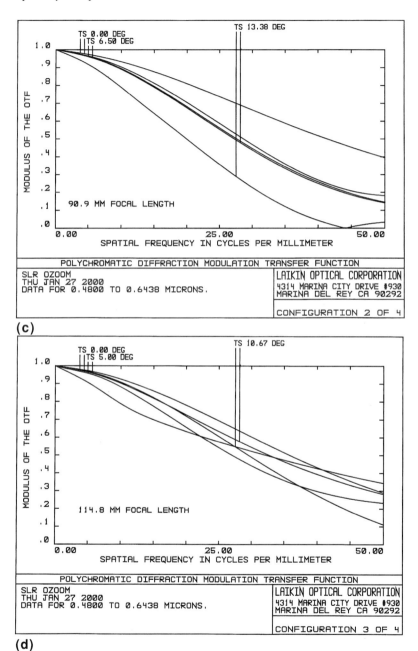

(d)

Figure 36-2 Continued

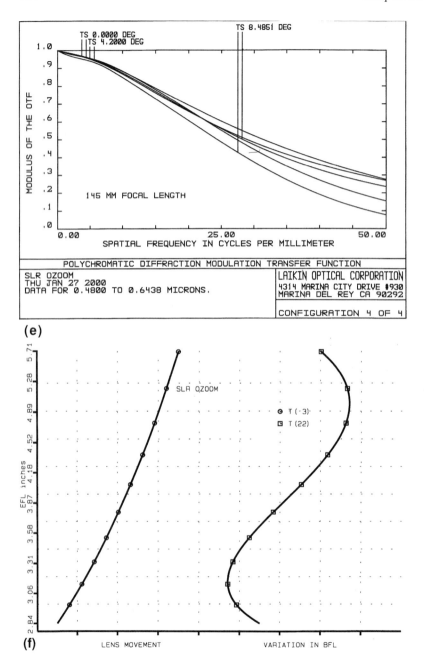

Figure 36-2 Continued

Optically Compensated Zoom Lenses

Table 36-3 72-145mm Zoom

Surf	Radius	Thickness	Material	Diameter
0	0.0000	1.0000E+10		
1	6.3722	0.2000	SF4	3.900
2	2.4419	1.4087	BAFN10	3.580
3	−11.3894	0.1950		3.580
4	−4.2529	0.1824	SK14	1.980
5	5.1930	0.2837		1.860
6	−3.8762	0.1614	SK14	1.860
7	1.8348	0.4068	SF4	1.920
8	7.4125	1.7731		1.920
9	−56.1977	0.4140	SK14	1.840
10	−3.4184	0.0193		1.920
11	3.5401	0.3230	SK14	1.840
12	−3.3877	0.0900	SF4	1.840
13	−27.6999	0.0223		1.840
14	Iris	0.0426		0.903
15	−2.6701	0.0750	BAK2	0.920
16	13.1555	0.0188		0.960
17	2.5097	0.1780	BAFN10	1.080
18	−4.4983	0.3888		1.080
19	−2.3194	0.0750	SF2	1.020
20	2.0389	0.1558		1.060
21	4.0040	0.2505	LAK9	1.280
22	−2.1983	3.5669		1.280

The distance from the first lens surface to the image is 10.543.

Table 36-4 Zoom Movements

EFL	T(3)	T(8)	T(13)	Distortion (%)	Pupil	Length
2.835	0.195	1.773	0.022	−6.20	3.688	10.231
3.580	0.640	1.328	0.467	−2.69	4.815	10.676
4.521	1.069	0.899	0.896	−0.27	6.289	11.105
5.709	1.489	0.479	1.316	1.41	8.343	11.525

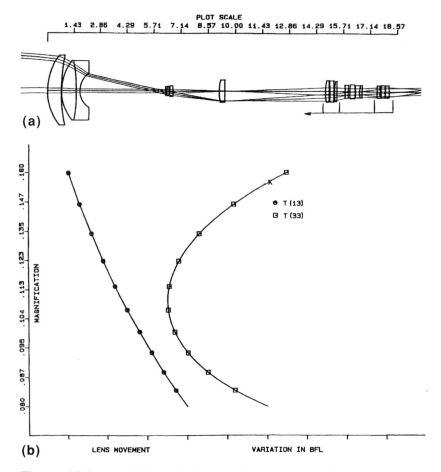

Figure 36-3 (a) 6.25 to 12.5× optically compensated projection lens. (b) Lens movement and variations in BFL. (c) 12.5×. (d) 9.921×. (e) 7.874×. (f) 6.25×.

screen to the first lens surface is 30 inches. Following is the lens prescription when in the 12.5× magnification position (Table 36-5).

The distance from the first lens surface to the image is 22.205. The entrance (as seen from the screen side) pupil distance is 17.657 and is a constant 0.4 diameter. Figure 36-3b shows lens movement and the variation of the paraxial location of the image vs. magnification.

Optically Compensated Zoom Lenses

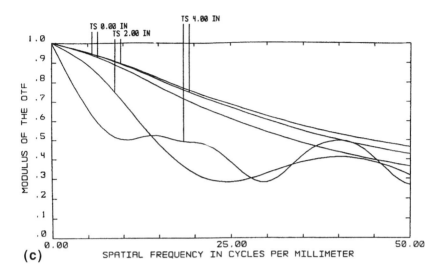

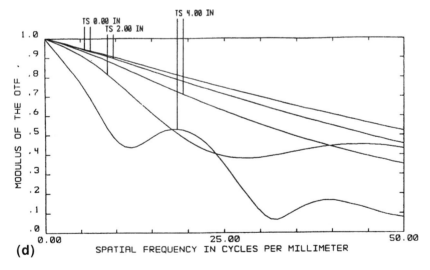

Figure 36-3 Continued

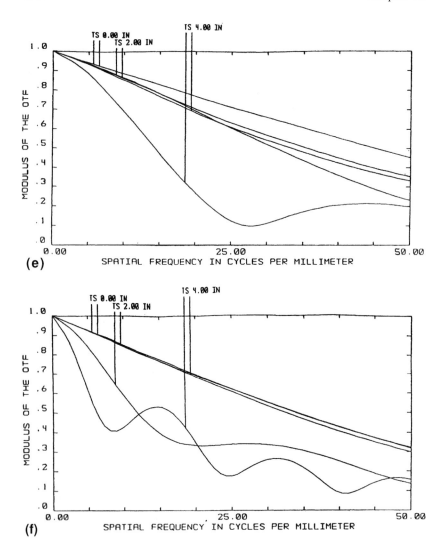

Figure 36-3 Continued

Table 36-6 Optically Compensated Projection Lens

Surf	Radius	Thickness	Material	Diameter
0	0.0000	30.0000		8.000
1	2.5446	0.7345	SK16	3.380
2	7.0456	0.0200		3.380
3	2.0978	0.8020	SK16	2.880
4	−7.5172	0.1900	LF5	2.880
5	0.9438	4.2107		1.660
6	Stop	0.3882		0.169
7	−0.9443	0.0500	LF5	0.340
8	0.8194	0.1300	SK16	0.460
9	−1.7325	0.0220		0.460
10	1.8215	0.1800	SK16	0.520
11	−3.4941	2.4236		0.520
12	2.1220	0.3000	SSKN5	1.080
13	−18.9215	5.3285		1.080
14	2.3567	0.2040	SK16	1.040
15	−16.9190	0.0539		1.040
16	−4.3567	0.2000	F2	0.940
17	−12.8973	0.0224		1.040
18	2.1351	0.1408	SK16	0.980
19	−28.6302	0.4280		0.980
20	−2.6205	0.1500	SF1	0.700
21	1.7463	0.0932		0.660
22	2.0930	0.2918	LAKN22	0.680
23	−1.8006	0.0200		0.680
24	−2.0298	0.2000	LAK16A	0.620
25	0.8564	0.1324	F6	0.620
26	1.7744	0.8098		0.620
27	33.2129	0.2001	SK16	0.620
28	−1.9145	0.0208		0.620
29	9.8754	0.2000	F2	0.620
30	4.2642	0.0253		0.560
31	12.8548	0.2014	SK16	0.620
32	−2.4380	4.0315		0.620

Table 36-6 Zoom Movements

Mag.	T(13)	T(19)	T(26)	T(32)	Distortion(%)
12.5	5.328	0.428	0.810	4.031	−1.6
9.921	5.136	0.621	0.617	4.224	−1.2
7.874	4.945	0.812	0.426	4.415	−0.9
6.25	4.755	1.001	0.236	4.605	−0.6

(We are tracing from the screen to the "image") Note that although we have four variable air spaces, there are only two compensation positions. MTF data is shown in Figs. 36-3c through 36-3f. The main aberration is primary lateral color. Table 36-6 lists zoom movements and distortion vs. magnification.

REFERENCES

Back, F. (1963). Zoom projection lens, U. S. Patent #3094581.

Grey, D. (1974). Compact eight element zoom lens with optical compensation, U.S. Patent #3848968.

Hiroshi, T. (1983). Bright and compact optical compensation zoom, U.S. Patent #4377325.

Ikemori, K. (1980). Optically compensated zoom lens, U.S. Patent #4232942.

Macher, K. (1969). Optically compensated varifocal objective, U.S. Patent #3451743.

37
Copy Lenses with Variable Magnification

In a xerographic copying machine and similiar applications, the object and image are fixed. The lens is allowed to move to obtain various magnifications. In Fig. 37-1a such a copy lens is shown (high magnification position). The aperture stop (surface 10) is just to the left of the

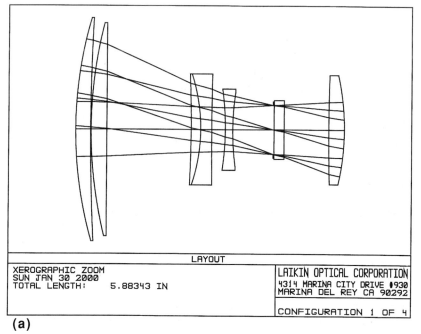

(a)

Figure 37-1 (a) Xerographic zoom lens. (b–e) Lens MTF.

Table 37-1 Xerographic Zoom Copy Lens (High Magnification Position)

Surf	Radius	Thickness	Material	Diameter
0	0.0000	21.2311		11.120
1	8.4566	0.3261	SK16	4.780
2	45.5618	0.0100		4.780
3	9.2944	0.3404	SK16	4.560
4	0.0000	1.8552		4.560
5	−24.4940	0.2116	SK16	2.380
6	−3.6662	0.2480	LF5	2.380
7	61.4561	0.2986		2.380
8	−5.8879	0.1653	LF5	1.740
9	5.3808	0.8823		1.580
10	Stop	0.0197		1.051
11	90.0804	0.2203	SK10	1.260
12	0.0000	0.9623		1.260
13	37.9841	0.3436	SK16	2.340
14	−5.2393	27.0195		2.340

next-to-last lens. It was designed to cover an image size of 13.9 diagonal, corresponding to a standard $8\frac{1}{2} \times 11$ inch paper. The object-to-image distance is fixed at 54.134 inch.

The variable air spaces are given in Table 37-2.

MTF data at the above magnifications is given in Fig. 37-1b through 37-1e. This lens system has considerable longitudinal chromatic aberration. At the high magnification position this is 1.27 mm (green to red).

Table 37-2 Variable Spaces and Distortion vs. Magnification

Mag	$T(0)$	$T(4)$	$T(12)$	$T(14)$	Distortion (%)
0.800	27.178	1.797	1.013	21.081	0.45
0.928	25.406	1.745	0.103	23.013	0.37
1.077	23.379	1.762		25.041	0.26
1.250	21.230	1.855		27.020	0.11

Copy Lenses with Variable Magnification 403

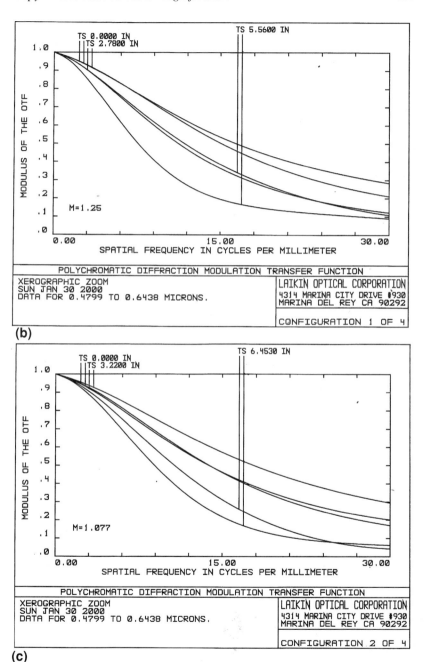

Figure 37-1 Continued

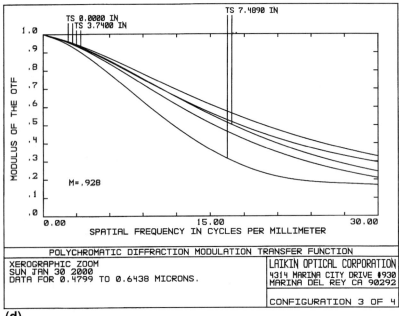

(d)

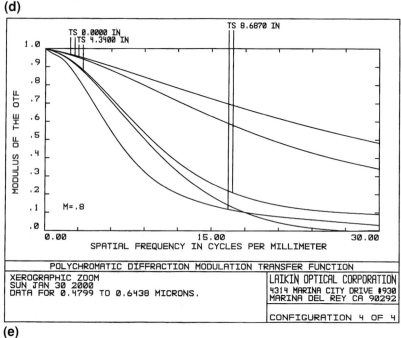

(e)

Figure 37-1 Continued

Copy Lenses with Variable Magnification

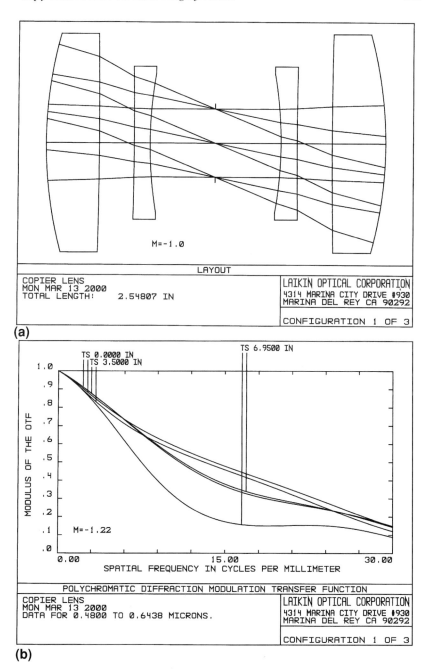

Figure 37-2 (a) Copier lens. (b–d) Lens MTF.

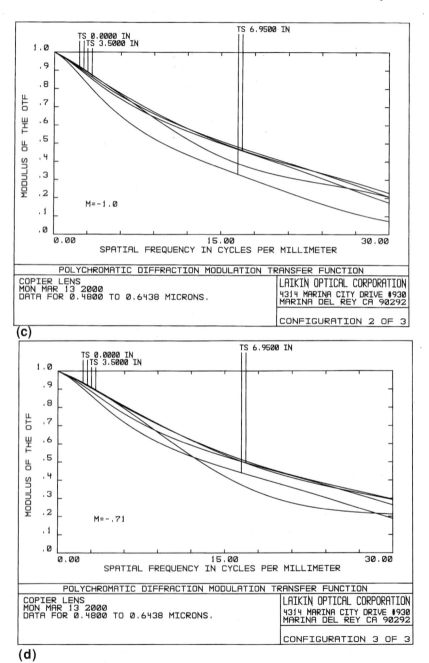

Figure 37-2 Continued

Table 37-3 Copier Lens

Surf	Radius	Thickness	Material	Diameter
0	0.0000	19.4935		13.900
1	2.9849	0.4000	LAKN12	1.600
2	−38.8801	0.2639		1.600
3	−27.5817	0.1200	LF5	1.120
4	2.9046	0.4901		1.020
5	Stop	0.4901		0.500
6	−2.9046	0.1200	LF5	1.020
7	27.5817	0.2639		1.120
8	38.8801	0.4000	LAKN12	1.600
9	−2.9849	23.9584		1.600
10	0.0000	0.0000		16.875

Table 37-4 Copier Lens

T(0)	T(4)	T(9)	Magnification	Distortion %
19.493	0.490	23.958	1.223	−0.04
21.591	0.720	21.402	1.0	0.0
26.157	0.054	18.166	0.7095	0.01

Note that this type of lens, although mechanically compensated in the sense that the moving groups operate under cam control, changes its overall length. (However, the object-to-image distance remains constant.)

In Fig. 37-2 is shown a variable magnification copy lens. It has a fixed object distance of 46 inch. It covers an object 14 inch diameter. The lens is symmetrical about the stop. It is a modification of the Yamakawa (1997) patent. Lens prescription is given in Table 37-3 and spacing data in Table 37-4.

REFERENCES

Arai, Y. and Minefuji N. (1989). Zoom lens for copying, U.S. Patent #4832465.

Harper, D. C., McCrobie, G. L., and Ritter, J. A. (1975). Zoom lens for fixed conjugates, U.S. Patent 3905685.

Yamakawa, H.(1997). Zoom lens system in finite conjugate distance, U.S. Patent #5671094.

38
Variable Focal Length Lenses

Lenses of this type predate true zoom lenses. It was realized some time ago that for a lens system consisting of positive and negative groups, varying the air space between these groups would change the effective focal length. For projection this is a handy feature since by a simple adjustment and then refocusing, the image may be made to fit the screen.

In Fig. 38-1a is shown a variable focal length lens suitable to

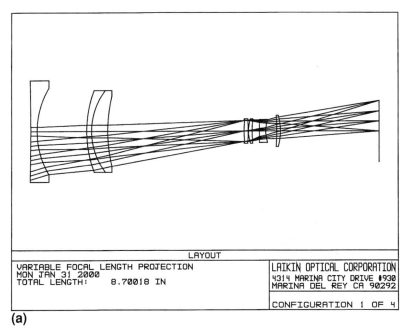

(a)

Figure 38-1 (a) Variable focal length projection lens. (b–e) Lens MTF.

409

410 Chapter 38

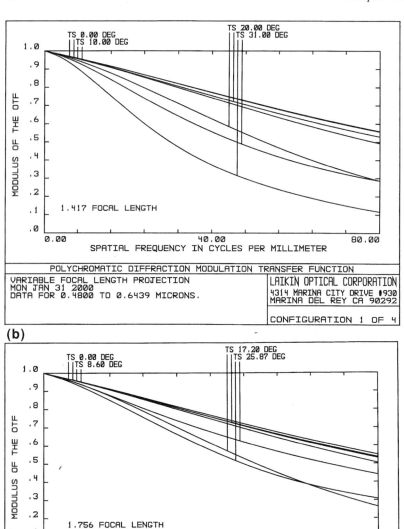

Figure 38-1 Continued

Variable Focal Length Lenses

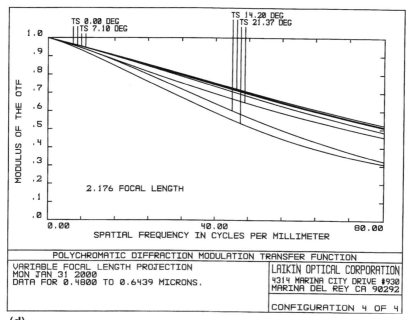

(d)

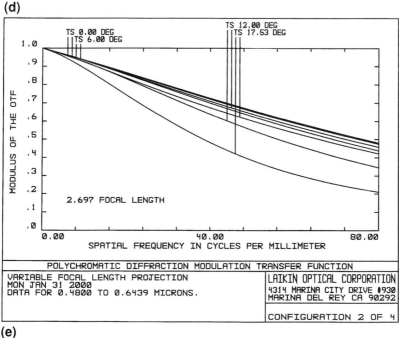

(e)

Figure 38-1 Continued

Table 38-1 Variable Focal Length Projection Lens

Surf	Radius	Thickness	Material	Diameter
0	0.0000	1.0000E+10		
1	−50.0181	0.1575	LAF2	2.480
2	2.0722	1.2396		2.120
3	2.8295	0.1181	LAFN7	2.000
4	1.6333	0.4022	SF55	2.000
5	3.6331	3.4241		1.720
6	Stop	0.0000		0.512
7	3.2202	0.1000	LAK10	0.600
8	−4.6843	0.0110		0.600
9	0.9301	0.0917	BAF51	0.600
10	3.0263	0.1816		0.600
11	−20.0034	0.1568	SF6	0.560
12	0.9297	0.2698		0.560
13	−5.2817	0.0901	BALF4	0.760
14	−1.4390	2.4578		0.760

Table 38-2 Spacings vs. Focal Length

Focal length	θ (deg)	Length	DIST. (%)	T(5)	BFL	f#
1.417	31.002	8.700	−12.25	3.424	2.458	6.71
1.756	25.870	7.764	−7.04	2.258	2.687	7.19
2.176	21.372	7.107	−3.69	1.317	2.972	7.78
2.697	17.526	6.700	−1.72	0.556	3.326	8.52

project film from a 35 mm SLR camera (18 × 24 mm format). In Table 38-1 is its lens prescription in the short focal length position, 1.417 (36.0 mm).

Figures 38-1b to 38-1e display the MTF at the above focal lengths. In Table 38-2, length is the length of the lens system, front lens vertex to image. Distortion is at full field. θ is the full field semifield angle. This lens system is a modification of the Sato (1988)

Variable Focal Length Lenses

patent. As with previous examples of projection lenses, an infinite object distance is used. It is recommended that a final computer run, at a typical finite conjugate, be made. Note that the stop is in contact with surface 7. The distortion is a little too large for most applications and so should be reduced.

In Fig. 38-2a is shown a variable focal length projection lens (see Angenieux, 1958) to project 35 mm film (1.069 diagonal). Unlike the above example, the length of the lens does not change. The inner lens group moves to vary the focal length from 1.772 (45 mm) to 2.165 (55 mm). It is f/3.44.

Table 38-3 lists the lens prescription (short focal length position).

Table 38-4 is a table of focal lengths vs. lens movements. This is

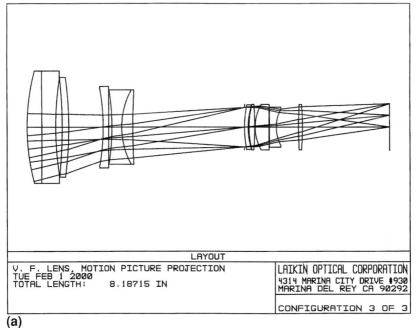

(a)

Figure 38-2 (a) Variable focal length motion picture projection lens. (b–e) Lens MTF.

Table 38-3 Variable Focal Length Projection Lens

Surf	Radius	Thickness	Material	Diameter
0	0.0000	1.0000E+10		
1	4.6897	0.3347	LAK9	2.480
2	−135.1888	0.3150	SF8	2.480
3	6.7937	0.1478		2.100
4	−9.9116	0.1500	LAF2	2.100
5	−7.8073	0.0421		2.200
6	−4.7155	0.1260	F5	1.720
7	−172.3283	0.0759		1.800
8	−4.9638	0.2247	LAKN12	1.640
9	1.6566	0.2524	SF8	1.640
10	14.7483	3.2415		1.640
11	Stop	0.0000		0.868
12	2.3852	0.0820	BASF2	0.980
13	1.5410	0.0200		0.900
14	1.9066	0.1290	LAKN12	1.000
15	−6.4898	0.0100		1.000
16	0.9935	0.3114	LAKN13	1.000
17	21.7701	0.0127		1.000
18	−37.7932	0.1547	SF8	0.880
19	0.8207	0.5040		0.800
20	8.0699	0.0891	LAKN13	1.000
21	−6.0327	1.8589		1.000

The distance from the first lens surface to the image is 8.187.

Table 38-4

EFL	T(5)	T(10)	BFL	Distortion (%)
1.772	0.042	3.242	1.859	−7.21
1.969	0.437	2.846	1.913	−5.57
2.165	0.780	2.503	1.964	−4.27

Variable Focal Length Lenses *415*

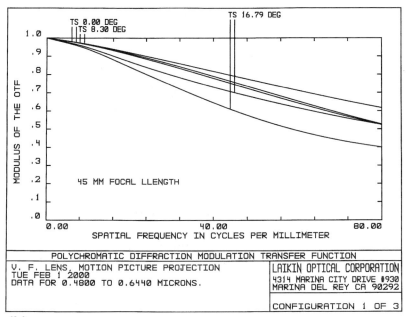

(b)

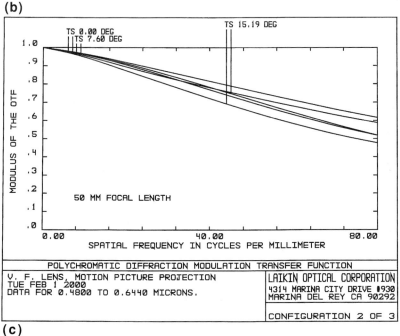

(c)

Figure 38-2 Continued

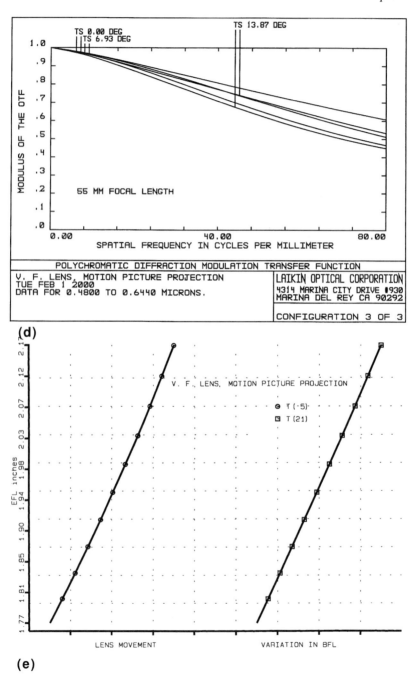

Figure 38-2 Continued

plotted in Fig. 38-2b. Notice that the back focal length increases with focal length. In the short focal length position, the exit pupil is located 2.929 to the left of the image surface. As discussed in Chap. 22, this is a reasonable distance for this 35 mm film projection. The stop is in contact with surface 12.

MTF data is shown in Figs. 38-2c through 38-2e.

REFERENCES

Angenieux, P. (1958). Variable focal length objective, U.S. Patent #2847907.
Sato, S. (1988). Zoom lens, U.S. Patent 4792215.

39
Gradient Index Lenses

In the previous chapters, we have assumed that the refractive index and dispersion of the medium were the same for all locations in the lens. Here we will discuss lens systems in which the refractive index has been made to vary depending upon its location in the lens. There are basically two types to consider, axial gradients and radial gradients.

AXIAL GRADIENT LENSES

In a lens with an axial gradient, the refractive index varies according to its position along the optical axis. A typical equation describing this is (Marchand, 1978).

$$n = n_0 + f(z)$$

where n_0 is the refractive index at the starting position (Z_0) and $f(z)$ is an index function referenced from this starting position. Ray tracing is accomplished by dividing the lens into small slices (ΔZ). As the ray passes through the lens, the value of Z is determined at each slice and the refractive index at that location computed. This is illustrated in Fig. 39-1. Large slices are taken for the preliminary system and reduced as the design progresses. Thus the designer has as a variable, for any given material, the value of T, the distance the lens vertex is from Z_0.

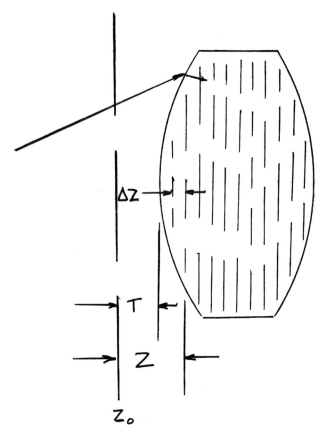

Figure 39-1 Tracing through a lens with an axial gradient.

This value of T must be indicated on the lens drawing. Unlike conventional glass blocks, this blank cannot be reversed by the optical shop.

ZEMAX, as well as most comprehensive lens design programs, provides the designer with several types of axial gradient functions. One such type is the GRADIUM™ surface (Lightpath, 1996). Figure 39-2 shows the index profile for one of their materials.

Gradient Index Lenses 421

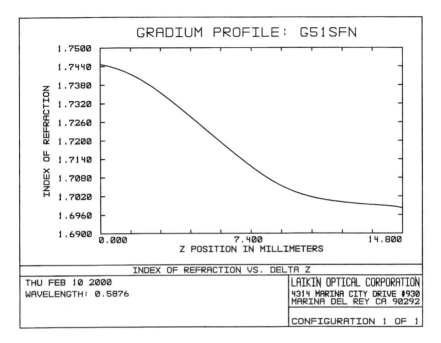

Figure 39-2 Index profile for GRADIUM™ surface.

In Fig. 39-3a is shown a 10× microscope objective using an axial gradient material. ΔZ is 0.0448. Figure 39-3b shows the MTF for this lens. Comparing to the design of Fig. 11-1, we note improved performance using an axial gradient. Magnification, numeric aperture, and field size is the same as in Fig. 11-1. Distortion is –0.14%.

Compare this to the design using a radial gradient material (Krishna, 1996).

In Fig. 39-4a is shown a lens of 6 inch focal length and f/4 with an axial gradient material. The field of view is 1.0 degree. Figure 39-4b shows the MTF for this. Note that positive spherical lenses using homogeneous glass have spherical aberration such that rays at the margin are focused closer to the lens than paraxial rays. The axial gradient solves this problem with a negative gradient. This causes the refractive

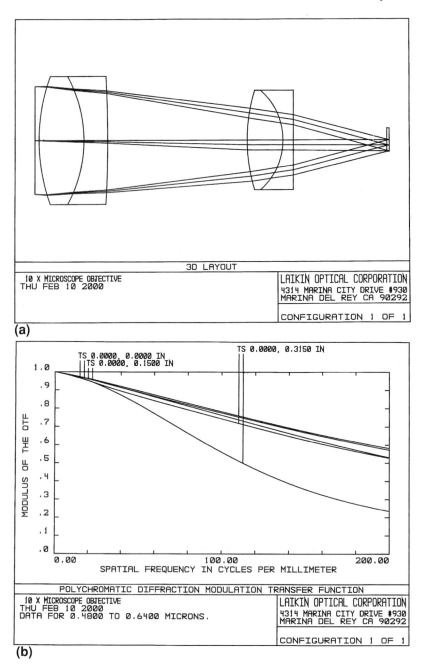

Figure 39-3 (a) 10× microscope objective lens. (b) Lens MTF.

Gradient Index Lenses 423

Table 39-1 10× Microscope Objective

Surf	Radius	Thickness	Material	Diameter
0	0.0000	6.0639		0.630
1	Stop	0.0120		0.303
2	0.5052	0.1309	K5	0.360
3	−0.3671	0.0700	F2	0.360
4	−3.2840	0.4006		0.360
5	0.4699	0.1023	BASF51	0.280
6	−0.1845	0.0300	G51SFN	0.280
7	0.0000	0.2700		0.280
8	0.0000	0.0070	K5	0.067
9	0.0000	0.0000		0.065

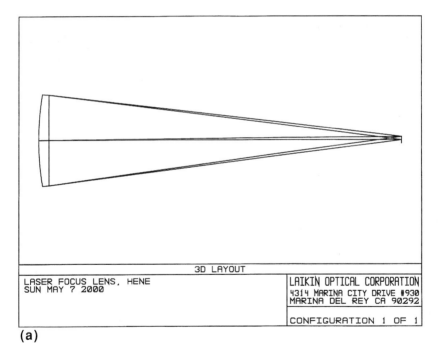

(a)

Figure 39-4 (a) 6-in. focal length, $f/4$ laser focusing lens. (b) Lens MTF.

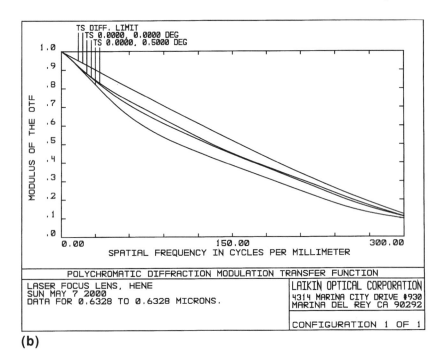

(b)

Figure 39-4 Continued

Table 39-2 6 Inch $f/4$, Laser Focus Lens

Surf	Radius	Thickness	Material	Diameter
0	0.0000	0.100000E+11		0.00
1	4.4701	0.1705	G23SFN	1.500
2	123.6921	5.9069		1.485
3	0.0000	0.0000		0.107

index to be less at the edge than at the center, thus reducing the spherical aberration.

ΔZ is 0. The reference index of refraction then is 1.7758. The second surface should be set plane for manufacturing reasons. The wavelength is 0.6328 μm.

RADIAL GRADIENT

A typical radial gradient has an index profile given by

$$n = n_0 + n_1 R^2$$

It is therefore possible to have a radial gradient material in the form of a plane plate and yet have power. Such a lens is listed in the following table. The MTF is plotted in Fig. 39-4. The effective focal length (at 0.55 μm) is 50. The material (radial) is per the above equation with $N_0 = 1.7$ and $N_1 = 0.05$. The MTF for this is shown in Fig. 39-5.

Long rod lenses, using a radial gradient index profile, can be made by an ion exchange process. Such devices are used in making

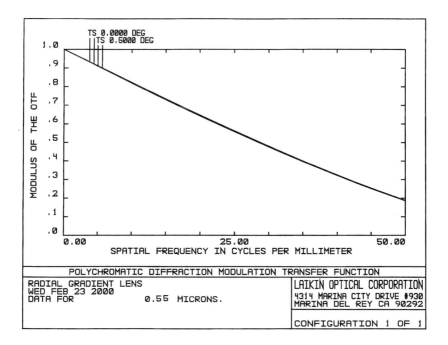

Figure 39-5 Radial gradient lens.

Table 39-3 Radial Gradient Lens

Surf	Radius	Thickness	Material	Diameter
0	0.0000	0.100000E+11		0.00
1	Stop	0.0000		2.000
2	0.0000	0.2000	Radial	2.000
3	0.0000	49.9592		2.000
4	0.0000	0.0000		0.874

long and very small diameter bore scopes for medical and industrial inspection purposes (Gradient Lens Corp., 1999). Such a device is shown in Fig. 39-6.

This has eight intermediate images inside the scope. This may be combined with a front objective to obtain an erect image when viewed with an eyepiece.

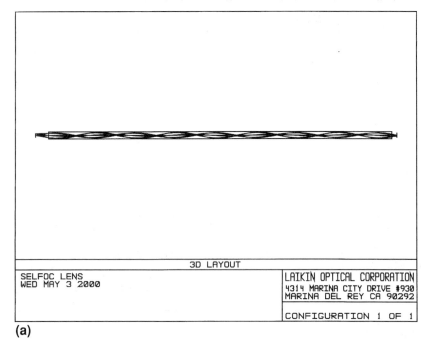

(a)

Figure 39-6 (a) Selfoc lens. (b) Lens MTF.

Gradient Index Lenses

Table 39-4 Selfoc Lens

Surf	Radius	Thickness	Material	Diameter
0	0.0000	0.0800		0.020
1	Stop	2.0546	SLS-1.0	0.041
2	0.0000	0.0298		0.041
3	0.0000	0.0000		0.024

The radial gradient material is made by NSG America (1999). Its index profile is given by

$$n = n_0 \left[1.0 - \frac{A}{2} r^2 \right]$$

where A is a function of the wavelength.

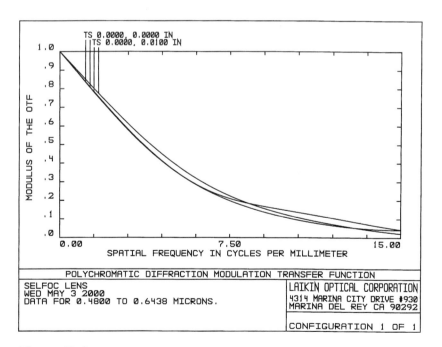

Figure 39-6 Continued

REFERENCES

Gradient Lens Corp., Rochester, NY (1999). Data sheet.

Greisukh, G. I. and Stepanov, S. A. (1998). Design of a cemented, radial gradient-index triplet, *Applied Optics, 37:*2687.

Krishna, K. S. R. and Sharma, A. (1996). Chromatic aberrations of radial gradient-index lenses I, *Applied Optics, 35:*1032.

Krishna, K. S. R. and Sharma, A, (1996). Chromatic aberrations of radial gradient-index lenses Π, *Applied Optics, 35:*1037.

Krishna, K. S. R. and Sharma, A. (1996). Low power gradient index microscope objective, *Applied Optics, 35:*5636.

Lightpath Technologies data sheet (1996). 6820 Academy Park East, NE, Albuquerque, NM 87109.

Manhart, P. K. (1997). Gradient refractive index lens elements, U.S. Patent #5617252.

Marchand, E. W. (1978). *Gradient Index Optics,* Academic Press, New York.

NSG America, Somerset, NJ (1999). Data sheet.

Rouke, J. L., Crawford, M. K., Fisher, D. J., Harkrider, C. J., Moore, D. T., and Tomkinson, T. H. (1998). Design of a three element night-vision goggle, *Applied Optics, 37:*622.

Appendix A
Film Formats

Camera	Dimensions (Diagonal)	Specification
8 mm	0.192 × 0.145 (.241)	PH22.19 [a]
SUPER 8	0.228 × 0.163 (.280)	PH22.157 [a]
16 mm	0.295 × 0.488 (.570)	
35 mm ACADEMY	0.866 × 0.630 (1.069)	PH22.59 [b]
ANAMORPHIC	0.864 × 0.732 (1.132)	PH22.59 [a]
FULL FRAME	0.980 × 0.735 (1.225)	PH22.59 [a]
35 mm SLR	1.417 × 0.945 (1.703)	RR-9-1966 [b]
VISTAVISION	1.486 × 0.992 (1.787)	
65 mm/70 mm	1.913 × 0.870 (2.101)	5 perf. PH22.152 [d]
870 format	1.913 × 1.434 (2.391)	8 perf.
10 Perf 70	1.913 × 1.808 (2.632)	
	2.799 × 2.072 (3.482)	15 perf Imax system PH22.145 [c]

[a] Standards are available from the American National Standards Institute, 430 Broadway, New York, NY 10018.
[b] Available from Society of Motion Picture and TV Engineers, 9 East 41 St., New York, NY 10017.
[c] Shaw, W. (1970). New large screen motion picture system, *SMPTE, 79:782*.
[d] Hecht, J. (1996). The amazing optical adventures of Todd-AO, *Optics and Photonics News*, 7:35.

Note: These are dimensions of the camera aperture; the aperture at the projector is slightly smaller. The values in parentheses correspond to the diagonal for a sharp corner aperture. However, the actual aperture plate has rounded corners, making the effective diagonal a little less.

In the Todd system of cinematography, 65 mm wide file is used. It is projected with 70 mm wide film. The extra width is used for multiple sound tracks.

ADDITIONAL FORMATS

$2/_3$ inch Vidicon	0.346 × 0.260 (.433)	Usually has a 1.5 mm face plate, $N=1.487$
$1/_2$ in. Vidicon	0.183 × 0.244 (.305)	Usually has a 1.5 mm face plate, $N=1.487$
1 in. Vidicon	0.5 × 0.375 (.625)	Usually has a 2.5 mm face plate, $N=1.487$
PLUMBICON	0.673 × 0.5 (.838)	
$1/_2$" CCD	0.252 × 0.165 (0.301)	

Appendix B
Flange Distances

Mount	Flange to image	Specification
C	0.690	PH22.76 [a]
T	2.169	
Arriflex	2.047 (52 mm PL mount)	
BNCR	2.420	
Panavision	2.030	
Nikon	1.831	
Olympus OM	1.819	
Hasselblad	2.828	
Mamiya Sekor	2.480	
35 and 70 mm projection	1.2 minimum	PH22.28 [a]

[a] American National Standards Institute, 1430 Broadway, New York, NY, 10018.

Appendix C
Thermal and Mechanical Properties

All values are at 20 degrees.

Material	Expansion coefficient ($10^{-6}/°C$)	$\Delta N/\Delta T$ ($10^{-6}/°C$)	Density (grams/cc)
As_2S_3	24.6		3.198
BaF_2	18.4	−15.2 (0.5 μm)	4.83
CaF_2	24	−11.2 (3.39 μm)	3.179
Germanium	6.1	403 (2.55 μm)	5.327
IRTRAN 1	10.2		3.18
IRTRAN 2	6.8		4.09
IRTRAN 3	18.7		3.18
IRTRAN 4	7.3	60 (10 μm)	5.27
IRTRAN 5	10.2		3.58
Lexan	67.5	−107 (0.58 μm)	1.20
MgF_2	13.1	1.9 (0.7 μm)	3.177
Pyrex	3.25		2.23
Plexiglas	67.5	−105 (0.58 μm)	1.18
Polystyrene	68		1.05
Quartz	0.52	10.0 (0.5 μm)	2.202
Sapphire	5.8	13 (0.55 μm)	3.98
Corning 7971	0.01	10.68 (0.5 μm)	2.205
Si	4.2	151 (2.55 μm)	2.329
ZnS	7.85	41 (10.6 μm)	4.09
ZnSe	7.57	61 (10.6 μm)	5.27
As_2S_3	21.4	9.3 (5 μm)	3.2
CdTe	5	9.8 (3.39 μm)	5.86
AMTIR-1	12	77 (3.39 μm)	4.4
AMTIR-3	13.5	92 (5 μm)	4.67
Gallium arsenide	6	216 (5 μm)	5.31
Xeonex	75	70	

REFERENCES

AMTIR-1 and AMTIR-3 (1993). Data sheets, Amorphous Materials, Garland, TX.

Corning (1985). Low expansion material, Corning booklet #7971.

Corning (1986). Fused silica, Corning booklet FS7490.

Dynasil (1988). Synthetic fused silica, Dynasil Catalog 302-M.

Feldman, A., Horowitz, D., and Walker, R. M., (1977). Optical materials characterization, AD A 045095.

Jacobs, S. F. (1987). Dimensional stability of materials useful in optical engineering, *Applied Optics and Optical Engineering*, Volume 10, 71, Academic Press, New York.

Moses, A. J. (1970). Refractive index of optical materials in the infrared, AD 704555, Clearinghouse for Federal Scientific and Technical Information, Springfield, VA.

Moses, A. J. (1971). *Handbook of Electronic Materials* Vol. 1, *Optical Materials Properties*, Plenum Press, New York.

Wolf, W. L., Platt, B. C., and Icenogle, W. H., (1976). Refractive index of germanium and silicon as a function of wavelength and temperature, *Applied Optics, 15:2348*.

Appendix D
Commercially Available Lens Design Programs

Engineering Calculations   (KDP)
1377 E. Windsor Rd. #317
Glendale, CA 91205

Gibson Optics   (OSDP)
655 Oneida Dr.
Sunnyvale, CA 94087

Gregory Optics   (OASYS)
Star Route A
Dripping Springs, TX 78620

Focus Software, Inc.   (ZEMAX)
P.O. Box 18228
Tucson, AZ 85731

Kidger Optics   (SIGMA)
9a High Street   (also includes lens library)
Crowborough, East Sussex TN6 2QA
England

Optical Research Associates   (Code V)
3280 Foothill Blvd. #300
Pasadena, CA 91107

Optical Systems Design (SYNOPSIS)
Farnham pt. P.O. Box 247
East Bombay, ME 04544

Optikos (ACCOS)
286 Cardinal Medeiros Ave.
Cambridge, MA 02141

Optis (SOLSTIS)
Toulon Cedex 9, France
USA Rep. Advanced Photonics
54 Plymouth Rd.
White Plains, NY 10603

Ray Cad (Auto Ray)
77 Scribner Road
Tyngsboro, MA 01879

Sciopt Enterprises (OPTEC)
P.O. Box 20637
San Jose, CA 95160

Sinclair Optics (OSLO and GENII)
6780 Palmyra Rd.
Fairport, NY 14450

Don Small Optics (SODA)
24271 Verde St.
El Toro, CA 92630

Stellar Softrware (BEAM 4)
P.O. Box 10183
Berkeley, CA 94709

Technical Software (EZ-RAY) ray-trace and
3438 Woodstock Lane   analysis only
Mountain View, CA 94040

Appendix E
Program Optics

The disk contains several programs which should be useful in optical design.

The file OPTICS.EXE is an executable program and is ready to run, as is, on an IBM compatible machine with at least WINDOWS 95® (note, since this is a 32 bit program, it will not run under DOS or WINDOWS 3.x). Also on this disk are the following data files:

1. GLASS, Schott glass catalog and some miscellaneous materials (water, etc.)
2. IRGLAS, some infrared materials
3. HOYA, Hoya glass catalog
4. OHARA, Ohara glass catalog
5. GALLIUM, gallium arsenide data to test program Polynomial Fit
6. BK-7, BK-7 optical glass data to test program Polynomial Fit
7. LENSCAM, example data for program Zoom Cam
8. LENSZ, data for program Zoom Spacings, a 10× mechanically compensated lens
9. LS, data file for the Drawing program.
10. DRAWFILE, common data entries for the Drawing program
11. PETZVAL, File for the Test Glass Fitting program
12. Test plate lists for
    HJGLAS, Harold Johnson Optical Lab
    COASTAL, Coastal Optical Systems

MODEL, Model Optics
OPTIMAX, Optimax
OGF, Optical Glass Fabricators
OCL, Optical Components Incorporated

In all these programs, data from the keyboard should be input with a decimal point (except surface numbers and other integer items) and separated with a comma.

To install these files from the START MENU,
RUN
A:INSTALL

A new directory will be created (OPTICS) and all the files installed in this directory.

Data plotting is done in the Hewlett Packard HP-GL/2 language (a vector graphics language) and thus should plot properly on any Hewlet Packard Laserjet printer. For proper plotting, the system should be in inches.

PROGRAM ACHROMAT

This program calculates the curvatures required to yield a thin lens achromat, for an object at infinity, of the desired focal length. Third-order coma and spherical aberration, as well as longitudinal chromatic (primary) aberration, is set to zero. The equations used are given on pp. 45 and 46.

The user is prompted to input the index of refraction and V values for the two elements (the A lens and the B lens) as well as the desired effective focal length.

The four required curvatures are then output along with PC, the Petzval sum. This is the curvature of the image surface in the absence of astigmatism.

Following are three examples, all of focal length 10.

BAK2/F2
BK7/F2
BAK1/SF8

Comparing C2 and C3, note that for the last case, the difference between C2 and C3 is the smallest; so it would be the best choice for a cemented achromat.

Program Achromat, Thin Lens

N1	V1	N2	V2	C1	C2	C3	C4	EFL	PC
1.5421	59.44	1.6241	36.11	0.1610	−0.3090	−0.3048	−0.0568	10.0000	0.0699
1.5187	63.96	1.6241	36.11	0.1646	−02781	−0.2749	−0.0671	10.0000	0.0714
1.5749	57.27	1.6942	30.95	0.1566	−0.2219	−0.2217	−0.0523	10.0000	0.0688

ACHROMATIC PRISM

This program calculates prism angles to make an achromatic prism of the desired deviation. Refer to p. 288.

The user is prompted to input index of refraction data for the two prisms and the angle of incidence. (The case for zero angle of incidence is shown in Figure 26-1 and Table 16-1.) Since ten values of deviation are computed, the user is prompted to input the smallest value and the increment.

Data may either be displayed on the screen or printed.

The system is achromatic if the deviation is the same at the extreme wavelengths. Desired deviation is for the central wavelength. Secondary is the deviation corresponding to the short wavelength minus the deviation at the central wavelength expressed in radians. Magnification is diameter of input beam/diameter of output beam.

The following example is for BAK1/F2. Input data is

1.56997,1.57487,1.58
1.61582,1.62408,1.6331

ACHROMATIC PRISM PROGRAM
INDEX AND V, PRISM 1 AND 2   1.57487 57.315 1.62408 36.116
ANGLE OF INCIDENCE = 0.000 DEGREES

Deviation	Angle 1	Angle 2	Secondary	Mag
0.500	2.350	1.364	−0.000004	1.000161
1.000	4.691	2.720	−0.000009	1.000642
1.500	7.016	4.061	−0.000013	1.001446
2.000	9.315	5.380	−0.000017	1.002573
2.500	11.581	6.670	−0.000022	1.004025
3.000	13.807	7.926	−0.000026	1.005803
3.500	15.987	9.142	−0.000030	1.007912
4.000	18.116	10.315	−0.000035	1.010353
4.500	20.187	11.439	−0.000039	1.013181
5.000	22.202	12.512	−0.000044	1.016249

BEST FORM LENS

This program calculates the lens shape for minimum third-order spherical aberration for an object at infinity. The equation used is given on p. 212.

The user is prompted for the desired effective focal length, lens diameter, index of refraction, and edge thickness. The program solves, by an iterative scheme, a thick lens with the above properties.

In output the radii, center thickness, and back focal length are listed.

The following shows two cases; first is F2 glass(N_e=1.62408), the second BK7(N_e=1.51872). Edge = 0.1.

PROGRAM BEST

Minimum Spherical Aberration Lenses

F	R1	R2	TH	DIA	BFL
10.000	6.551	−129.579	0.283	3.000	9.834
10.000	5.939	−40.210	0.321	3.000	9.816

DESIGN A CONDENSER

This short program calculates the prescription for a symmetric condenser consisting of two plano-convex lenses separated by an axial distance of D. The user is requested to enter the lens effective focal lenth EFL, lens diameter DIA, edge thickness E, and refractive index of the lenses.

The following data is output:

$D = E/2$.
$R =$ lens thickness.
$T =$ lens thickness.
$L =$ disance first surface to last $= 2T + D$.
$S =$ spacer thickness assuming that the inside diameter of this spacer is DIA - 2E.
H is the principal plane location as measured from the first plane surface.

See Fig. 29-1 in Lens Design.

The following is an example for a fused quartz condenser assembly.

****PROGRAM CONDENSER****

EFL	DIA		EDGE		INDEX
5.00000	3.00000		0.10000		1.46080
R	T	D	S	L	H
4.5965	0.3516	0.0500	0.4868	0.7523	0.2658

PROGRAM CUTTER

This program computes the X, Y coordinates for a conic section or general aspheric. It also computes the coordinates of a cutter to make this section. The aspheric equation is given on page 6. Referring to Fig. 1-1 on page 7, X, Y are the section coordinates; U, V are the cutter coordinates: and, ϕ is the slope of the tangent to the surface being generated. The equations used are given on pp. 7 and 8.

The paraxial radius of curvture is requested as well as the aspheric coefficients. Input all values in decimal form with due regard to sign convention. (If a negative value for the radious is input, the program will reverse the signs of the radius and the aspheric coefficients, not the conic coefficient.) Separate each value with a comma. If a concave section is to be generated, input the cutter diameter as a positive number. Likewise a negative cutter diameter will generate a convex section. The section may be plotted on the printer if requested. It will be scaled to fit the page.

For output, section values of X, Y, RADIUS, and SLOPE are printed. SLOPE is given in degrees, minutes, and seconds. Coordinates of the cutter U, V to generate this curve are also listed. Due to the large amount of data created, this data is not displayed but is instead always printed. The printer then must be on.

Aspheric Coordinate Data for Test Case for Aspheric Axial Radius = 6.00000000
Conic Coefficient = −1.00000000
Cutter Diameter = 0.50000
Aspheric Coefficients
0.10000000D-04   0.20000000D-04   0.30000000D-04
0.40000000D-04
A concave section is being generated

Section Values				Cutter Values	
X	Y	RADIUS	SLOPE	U	V
0.00000	0.00000	6.00000	90 0 0	0.00000	0.00000
0.00021	0.05000	6.00020	89 31 21	0.00020	0.04792
0.00083	0.10000	6.00082	89 2 43	0.00080	0.09583
0.00188	0.15000	6.00184	88 34 4	0.00180	0.14375
0.00333	0.20000	6.00327	88 5 27	0.00319	0.19167
0.00521	0.25000	6.00510	87 36 50	0.00499	0.23959
0.00750	0.30000	6.00732	87 8 15	0.00719	0.28752
0.01021	0.35000	6.00994	86 39 41	0.00978	0.33544
0.01333	0.40000	6.01293	86 11 8	0.01278	0.38337
0.01688	0.45000	6.01629	85 42 38	0.01618	0.43130
0.02083	0.50000	6.01998	85 14 9	0.01997	0.47924
0.02521	0.55000	6.02397	84 45 41	0.02417	0.52717
0.03000	0.60000	6.02821	84 17 16	0.02876	0.57512
0.03521	0.65000	6.03263	83 48 52	0.03376	0.62306
0.04084	0.70000	6.03713	83 20 30	0.03915	0.67101
0.04689	0.75000	6.04158	82 52 8	0.04495	0.71897
0.05335	0.80000	6.04580	82 23 46	0.05115	0.76692

	Section Values			Cutter Values	
X	Y	RADIUS	SLOPE	U	V
0.06024	0.85000	6.04955	81 55 22	0.05776	0.81487
0.06754	0.90000	6.05253	81 26 55	0.06476	0.86283
0.07528	0.95000	6.05435	80 58 20	0.07218	0.91077
0.08343	1.00000	6.05449	80 29 35	0.08000	0.95871
0.09202	1.05000	6.05234	80 0 34	0.08823	1.00663
0.10105	1.10000	6.04713	79 31 9	0.09688	1.05452
0.11053	1.15000	6.03792	79 1 13	0.10595	1.10238
0.12046	1.20000	6.02359	78 30 32	0.11545	1.15020
0.13086	1.25000	6.00283	77 58 52	0.12538	1.19794
0.14175	1.30000	5.97412	77 25 54	0.13576	1.24560
0.15316	1.35000	5.93577	76 51 14	0.14661	1.29314
0.17767	1.45000	5.82269	75 34 49	0.16979	1.38774
0.19085	1.50000	5.74410	74 51 45	0.18218	1.43472
0.20474	1.55000	5.64837	74 4 22	0.19515	1.48140
0.21942	1.60000	5.53402	73 11 41	0.20874	1.52772
0.23498	1.65000	5.40009	72 12 31	0.22303	1.57361
0.25156	1.70000	5.24633	71 5 34	0.23807	1.61899
0.26929	1.75000	5.07342	69 49 20	0.25395	1.66377
0.28837	1.80000	4.88315	68 22 11	0.27077	1.70785
0.17767	1.85000	4.67845	66 42 26	0.28865	1.75114

DRAWING PROGRAM

This program will make lens, mirror, and some mechanical part drawings suitable for submission to an optical or machine shop. A title box is provided, shop notes are added from a standardized list, the lens, mirror, or mechanical part is drawn, and a box is provided for radius, accuracy, irregularity, quality, and clear aperture data. Data may be input from a file or from the keyboard.

The user is first asked if the input prescription is in a file (see file LS included on the disk). This file has a title as the first line, a blank line, next a heading line (Radius, Thickness, etc.), another blank line, followed by KSUR + 2 lines, where KSUR is the numer of lens surfaces. The first line is the object surface and the last is the image surface, so they are not used by this drawing program. There are four columns of numbers; Radius, Thickness, Material, and Diameter. The example file is in fixed format; 14, 4X, F13.5, F13.4, A18, F8.3. The easiest way to create this file is simply to separate the data items with a comma. The first integer numbers on each line are for reference only; the program ignores them.

Note: The lens files provided on a separate disk are compatible with the above format.

Drawfile (also included on the disk) is a file containing the most common data entries. The first line has the company name, then the drawing title. Data is in lines 1–20. In column 1 on the next four lines are codes for coating type, wavelength region, glass material, and if there are cylinder or aspheric surfaces. Explanation of the codes is contained in columns 21–70 on each line of this sample file.

The user is next asked if the input data is in millimeters or inches. Then the following choices are presented.

1. Singlet lens
2. Doublet lens (see Fig. A1)
3. First surface mirror
4. Rear retainer (has a slot and a thread on the OD)
5. Simple spacer
6. Stepped spacer (see Fig. A2)
7. Lens caps (front and rear)
8. Front retainer (engraving on the front)
9. Thin spacer (brass shim stock)

The user is then asked if legal size paper (8.5 × 14 inch) is to be used. Legal size is recommended for cylinder lenses due to the extra view required.

When data is input from a lens file, the user is asked for the number of the first lens surface.

Whether lens data is input from a file or from the keyboard, Radius, Thickness, and Diameters are displayed. The user is asked to verify them, and they may be corrected at the keyboard.

Refer to pp. 27 through 30 for a discussion of drawing preparation.

For singlet, doublet, and mirror drawings, there is a box in the upper right hand corner for the user to enter (manually) the various tolerances on radius, accuracy, and irregularity. It is also necessary to indicate surface quality and surface clear apertures. It is within these clear apertures that the tolerances apply. Lenses should in general be at least 1 mm larger than the clear apertures.

Surface diameters and lens thicknesses are indicated on the drawing itself. Tolerances should be manually inserted on the drawing. Also the user needs to add to

> Note 1, the amount of break on the lens edge
> Note 6, the effective focal length (this is handy to check if the part has been properly made)

In the case of a cemented double add to

> Note 10, materials for the lens elements
> Note 12, the maximum power difference at the cemented surface

For a single-element lens, if both surfaces are concave, the user is asked whether the lens diameter is to be extended. This is sometimes convenient for mounting purposes. In the case of a cylinder, the user is asked for the width of the lens (the non-power dimension). For an aspheric, the user is asked to input the coefficients.

ABC Corporation (file LS)

	Radius	Thickness	Material	Diameter
0		0.10E+11		
1	17.68400	0.30000	FK5	2.400
2	−8.00100	0.30000	LAF2	2.400
3	2.32100	0.4123		1.940
4	Stop	0.30000		1.000
5	−2.03460	0.1422	F2	1.480
6	3.10910	1.0793		1.280
7	1.05635	0.6141	LAKN12	1.240
8	−2.09300	0.1464		1.240
9	−0.93500	0.1116	SF54	0.800
10	7.10500	0.4023		0.800

ABC CORPORATION	COMPANY NAME 20 CHARACTERS
400 mm projection	DRAWING TITLE, 20 CHARACTERS
1	1 FOR HEA COATING 2 MGF2 3 LASER V COATING
1	1 FOR VISUAL REGION 0 TO INPUT DATA IN MICRONS
1	1 1 FOR STANDARD GLASS MATERIAL, 2 TO INPUT TYPE
0	1 FOR CYLINDER/ASPHERIC 0 IF NONE

(DRAWFILE)

The following drawing, Fig. E1, was made using the above files. It is a cemented doublet; surfaces 1–3 of file LS Tolerances, accuracy, irregularity, etc., must be added by the user.

In Fig. E2 is shown a stepped spacer drawing made with this pro-

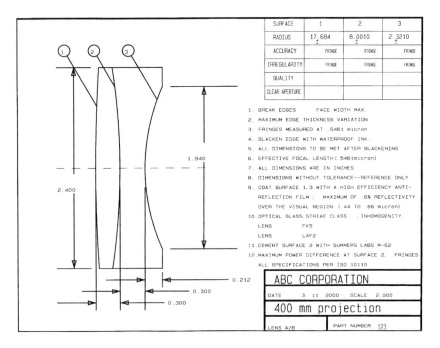

Figure E-1 400 mm projection lens.

Program Optics 447

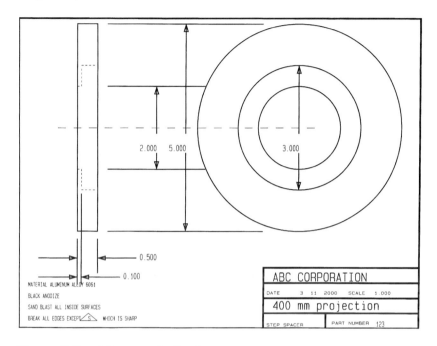

Figure E-2 400 mm projection lens.

gram. It has a step diameter of 2.0 and an inside diameter of 3.0. Tolerances must be added.

PROGRAM MIRROR

This program computes the primary and secondary mirror radii for a Cassegrain mirror system. The equations used are given on p. 182. Referring to Fig. 15-1, B, the back focal distance, is equal to $H\,F$.

The user is prompted for the desired effective focal length (F) and the paraxial axial height ratio (H).

Output (which may be printed if requested) consists of a table of primary (RP) and secondary (RS) mirror radii with the mirror spacing (T). The user should then pick a system appropriate to his needs. The Petzval sum will be zero when primary and secondary mirrors have the same radius.

Program Mirror
Cassegrain Lens System of Focal Length 100.00; $H = 0.300$

RP	RS	T
−160.00000	−240.00002	56.00000
−159.73332	−237.73534	55.93463
−159.28063	−234.15198	55.80401
−158.64320	−229.35145	55.60849
−157.82295	−223.46582	55.34865
−156.82244	−216.64893	55.02528
−155.64484	−209.06703	54.63941
−154.29394	−200.88974	54.19228
−152.77413	−192.28224	53.68532
−151.09032	−183.39905	53.12019
−149.24799	−174.37957	52.49871
−147.25313	−165.34547	51.82292
−145.11218	−156.39959	51.09500
−142.83203	−147.62605	50.31731
−140.41997	−139.09129	49.49233
−137.88367	−130.84570	48.62270
−135.23110	−122.92557	47.71116
−132.47051	−115.35510	46.76058
−129.61040	−108.14848	45.77388

PROGRAM MTF

This short program lists the diffraction-limited MTF response for a uniformly illuminated entrance pupil in incoherent light with no obscuration. The equation used is given on p. 38.

The user is prompted for the wavelength (in microns), f# of the lens, and the lp/mm increments. A table of response vs. lp/mm is then displayed (or printed if requested).

Program Optics

0.500 Microns, f/2.000

lp/mm	Response
10	0.987
20	0.975
30	0.962
40	0.949
50	0.936
60	0.924
70	0.911
80	0.898
90	0.886
100	0.873
110	0.860
120	0.848
130	0.835
140	0.822
150	0.810
160	0.797
170	0.785
180	0.772
190	0.760
200	0.747

POYNOMIAL FIT

This program fits index of refraction data, by a least squares fit, to a polynomial of the form

$$N_1^2 - 1 = \sum_{J=1}^{6} A_J \lambda_I^{2(L-J)}$$

For $L = 2$ we obtain coefficients for the Schott equation (NEQ = 0 in the Refraction program).

$$N^2 = F_1 + F_2\lambda^2 + F_3\lambda^{-2} + F_4\lambda^{-4} + F_5\lambda + F_6\lambda$$

For $L = 3$ we obtain coefficients for the polynomial equation (NEQ = 3 in the refraction program).

$$N^2 = F_3 + F_1\lambda^4 + F_2\lambda^2 + f_5\lambda^{-4} + F_6\lambda^{-6}$$

The user is first asked if the data is to be printed, then asked if data is to be read from a file. The first line of this file should be a title, 40 columns max. Next follow lines containing the wavelength in microns and the corresponding index of refraction. Separate values with a comma; write only one set of wavelength and index values per line (see file Poly).

If data is not to be read from a file, the user is asked for a title and then the number of items to be fitted (maximum 99).

Then follow wavelength and refractive index values (separated by comas).

Output are the coefficients in correct order for the Refractive Index program. RMS error is calculated for the fit. Pick the best fit ($L=2$ or 3 case) and insert this data into your appropriate glass catalog.

The following example is a fit for gallium arsenide (See file Gallium). As can be seen, $L=3$ is a better fit than $L=2$. These are the values used in Irglas. (pp. 453 and 456). Data is from the Amorphous Materials (Garland, TX) data sheet.

Program Polynomial
Gallium Arsenide
L = 2

F_1	F_2	F_3	F_4	F_5	F_6
1.093993D+01	−1.31164D+00	3.696380D+02	1.854296D−03	4.689746D+01	9.988722D+02

	Lambda	N (input)	N (calc.)	Error
1	2.500000	3.325600	3.325603	−0.000003
2	3.000000	3.316900	3.316878	0.000022
3	4.000000	3.306900	3.307034	−0.000134
4	5.000000	3.301000	3.300764	0.000236
5	6.000000	3.296300	3.296301	−0.000001
6	7.000000	3.292300	3.292243	0.000057
7	8.000000	3.287800	3.287987	−0.000187
8	9.000000	3.283000	3.283293	−0.000293
9	10.000000	3.277800	3.278065	−0.000265
10	11.000000	3.272500	3.272265	0.000235
11	12.000000	3.266600	3.265875	0.000725
12	13.000000	3.258900	3.258889	0.000011
13	14.000000	3.250900	3.251302	−0.000402

SUMSQ = 1.011144D−06.   RMS = 2.788911D−04.

PROJECTION LENS

This program computes the required effective focal lengths, film tilt, and displacement values as discussed in Chap. 22 (see Fig. 22-2 on p. 253) for a projection lens.

The program requires that the distance from the lens front principal plane to the screen be input. Distortion at the edge of the screen should be estimated.

If all linear dimensions are input in inches, the lens EFL will be given in inches as well as in mm. Tilt angles are in degrees.

Using first-order data, the required lens focal length will be computed. If the screen is tilted, the film tilt will be computed to maintain top and bottom in focus. This is the Scheimpflung condition.

If the lens is displaced from the screen center, the film shift is given to avoid keystone effects.

Screen Radius =	900.00000
Distortion =	3.00000%
Screen To Lens Distance =	900.00000
Film Width =	1.91300
Film Height =	0.86800
Screen Width at Chord =	300.00000
Height Lens is Above Screen Center =	0.00000
Horizontal Lens Displacement =	0.00000
Screen Tilt (degrees) =	0.00000
Lens Focal Length =	5.796 inches =147.22 mm
Lens Covers a Diagonal of	2.101 inches

$L = 3$

F_1	F_2	F_3	F_4	F_5	F_6
−1.743373D−06	1.087572D+01	1.409535D+00	−1.260622D−03	−2.116871D+00	5.019162D+00

	Lambda	N (input)	N (calc.)	Error
1	2.500000	3.325600	3.325607	−0.000007
2	3.000000	3.316900	3.316866	0.000034
3	4.000000	3.306900	3.306986	−0.000086
4	5.000000	3.301000	3.300974	0.000026
5	6.000000	3.296300	3.296317	−0.000017
6	7.000000	3.292300	3.292064	0.000236
7	8.000000	3.287800	3.287769	0.000031
8	9.000000	3.283000	3.283178	−0.000178
9	10.000000	3.277800	3.278126	−0.000326
10	11.000000	3.272500	3.272486	0.000014
11	12.000000	3.266600	3.266148	0.000452
12	13.000000	3.258900	3.259010	−0.000110
13	14.000000	3.250900	3.250969	−0.000069

SUMSQ = 4.249852D−07.   RMS = 1.808070D−04.

REFRACTIVE INDEX

This is an index of refraction interpolation program. Data for the entire Schott, Ohara and Hoya glass catalogs, as well as several other optical materials germanium, silicon, quartz, water, diamond, MgF_2, etc.) are stored in data files. Data is stored in any of four different types of interpolation equations. Index of refraction and dispersion data are computed for three to six wavelengths in microns.

The names of the data files are

Glass   contains the Schott glass catalog as well as some additional materials such as water, quartz, Plexiglas, calcium fluoride, etc.

Ohara   Ohara glass catalog

Hoya   Hoya glass catalog

Irglas   Various materials for the infrared region such as germanium, silicon, etc. You will be prompted for the name of the file containing the refractive index data. Data in these files are arranged as follows,

Eight places are allowed for the glass type followed by the six-digit glass code,

F_1 F_2 F_3 F_4 F_5 F_6 NEQ

NEQ	integer values in column 95
0	Or blank the Schott equation is used (L=2 in the Polynomial program)
1	Sellmeier equation
2	Sellmeier equation
3	Polynomial equation ($L = 3$ in Polynomial Fit)
4	Cauchy equation

Schott Equation

This equation was published in the Schott catalog some time ago. Originally the glass manufacturers listed these coefficients in their catalogs. The equation is (see p. 9)

$$N^2 = F_1 + F_2\lambda^2 + F_3\lambda^{-2} + F_4\lambda^{-4} + F_5\lambda + F_6\lambda$$

The glass catalogs generally list these coefficients as A0, A1, A2, etc.

Recently most glass manufacturers have fitted their data to a Sellmeier equation. This generally provides a more accurate fit of the data over a large wavelength region.

Sellmeier Equation

This is a very useful equation for most materials. Several researchers report dispersion data and then give the coefficients for the material. The equation used is

$$N^2 - 1 = \frac{F_1\lambda^2}{\lambda^2 - F_4} + \frac{F_2\lambda^2}{\lambda^2 - F_5} + \frac{F_3\lambda^2}{\lambda^2 - F_6}$$

Herzberger Equation

Dispersion data for several materials are given for an equation proposed by Herzberger. This equation is

$$N = F_1 + \frac{F_2}{\lambda^2 - F_6} + \frac{F_3}{(\lambda^2 - F_6)^2} + F_4\lambda^2 + F_5\lambda^4$$

This equation is generally valid only over a limited region. It thus must be used with caution. For most materials, $F_6 = 0.028$, corresponding to a strong absorption band at 0.167 μ $(\sqrt{.028} = 0.167)$

Polynomial Equation

$$N^2 = F_3 + F_1\lambda^4 + F_2\lambda^2 + F_4\lambda^{-2} + F_5\lambda + F_6\lambda\lambda_6$$

Cauchy Equation

$$N = A + \frac{B}{\lambda^2}$$

You are first prompted if you want the data printed. Next you are prompted to enter the wavelengths, in microns, in ascending order, with a maximum of six. Enter all values with a decimal point and separate with comas, as: 0.43583, 0.54607, 0.6328, 0.76819.

If three wavelengths are input, the output will be

Material N_1 N_2 N_3 V P

For glass materials, the glass type is listed followed by the six digit code per MIL-G-174. The first three digits are the refractive index at 0.5876 μm minus one, while the last three digits correspond to the Abbe *V* value without the decimal point.

$$V = \frac{N_2 - 1.0}{N_1 - N_3} \qquad P = \frac{N_1 - N_2}{N_1 - N_3}$$

If four, five, or six wavelengths are input, then only index of refracton data is output.

You will be prompted if computation of only one material is desired. If so, enter the material type (maximum of 8 spaces).

Printing is done with 57 lines per page. When printing is not requested, it is displayed 24 lines on the screen. The screen is not scrolled; press enter to see the next screen. Should you want to print subsequent data, strike 1 then Enter. This is useful if you only want the data for a few materials. Some comments regarding the data files:

GLASS

Quartz	Fused quartz or fused silica
Diamond	Data for diamond. Only valid for wavelengths less than 0.7μm
Plexiglas	Data for the acrylic material Plexiglas (Rohm and Haas) or Lucite (Dupont)
Polystyrene	Data for polystyrene available from Arco Polymers and Monsanto (Lustrex trade name)
Water	Data for pure distilled water
Sea Water	Data for typical ocean water (note the blank space)
CR-39	A thermosetting casting resin available from Pittsburg Plate Glass Industries
Zeonex	A polyolefin resin, distributed by BF Goodrich
Ultran30	Data for a new ultraviolet transmitting glass
Pyrex	Trade name of Corning Glass
Lexan	Trade name of General Electric for their polycarbonite resin
IRGLAS	
IRG-2, etc.	Infrared transmitting glasses

Strontium	Data for strontium titanate. This data is only valid for wavelengths greater than 0.8 μm
LiF	Lithium fluoride. Data only valid for wavelengths greater than 1 μm
Irtrn1, etc.	Sintered infrared materials
Ti Cha	Data for infrared transmitting chalocenide glass
Gallium Arsenide	Amorphous materials, Garland TX, data sheet

REFERENCES

Glass

Hoya Glass (1996). San Jose, CA (has a glass catalog program available).
Ohara Glass (1996). Somerville, NJ.
Schott Glass Tech. (1996). Durea, PA. (Has a glass catalog program available.)
Schott Glass Tech. (1999). Optical Glass, Product Range 2000.
Schott data sheet #3112e, Infrared Transmitting Glasses.

Quartz

Malitson, I. H. (1965). Interspecimen comparison of the refractive index of fused silica, *JOSA*, 55:1205.

Water

Centino, M. (1941). Refractive index of liquid water; *JOSA* 31:244.
Palmer, K. F. and Williams, D. (1947). Optical properties of water in the near IR, *JOSA*, 64:1107.

CaF_2

Malitson, I. H. (1963). A redetermination of some optical properties of calcium fluoride, *Applied Optics*, 2:1103.

BaF_2

Malitson, I. H. (1964). Refractive properties of barium fluoride, *JOSA*, 54:628.

Al$_2$O$_3$

Malitson, I. H. (1962). Refraction and dispersion of synthetic sapphire; *JOSA*, 52:1377.

Malitson, I. H. (1984). Optical properties and applications of sapphire, *Union Carbide Bulletin*, CPD7715.

Diamond

Wolf, W. L. (1978). Properties of Optical Materials, Sec. 7, Table 14, *Handbook of Optics* (Driscoll, W. G., ed.), Optical Society of America, McGraw-Hill, New York.

Plexiglas

Rhom and Haas Technical data sheet (1980). PL-165F.

Jans, R. W. (1979). Acrylic polymers for optical applications, *SPIE*, *204*: 2.

Lexan

General Electric Data Sheet (1983). CDC-683.

Walker, R. M., Horowittz, D., and Feldman, A. (1979). Optical and physical parameters of plexiglas and lexan, *Applied Optics*, 18:101.

IRG

IRG materials are glass materials with considerable infrared transmission. They are transparent from about 0.4 µm to about 4 µm (5.0 µm in the case of IRG11).

Schott data sheet, 3112e, Infrared-transmitting glasses.

IRG 2 Germanate glass
IRG 3 Lanthanum dense flint glass
IRG N6 CaAl silicate glass
IRG 7 Lead silicate glass
IRG 9 Fluorophosphate glass
IRG 11 Ca aluminate glass

IRTRAN

The IRTRAN materials are made by the Eastman Kodak Co. by hot-pressing the pure powders into an optical blank. They are then

sintered by inductive heating under very high pressures. Optical
Materials (1971). Kodak Publication U-72.

IRTRAN-1 MgF2
IRTRAN-2 ZnS
IRTRAN-3 CaF_2
IRTRAN-4 ZnSe
IRTRAN-5 Mg0

ZnS and ZnSe

Zns and ZnSe are made by a chemical vapor deposition process by Raytheon, II VI Inc., Phase4, and CVD Inc.

Ti

Ti chalcogenide is an infrared transmitting chalcogenide glass manufactured by Texas Instruments.

As_2S_3

As_2S_3 (arsenic trisulfide) is an infrared transmitting glass made by Amorphous Materials, Garland TX.

AMTIR-1 and AMTIR-3

Data sheet (1993). Amorphous Materials, Garland, TX.

IRG Glasses

Schott data sheet # 3112e (1992). Schott Glass Technologies, Durea, PA.

Si

Edwards, D. F. and Ochoa, E. (1980). Infrared refractive index of silicon, *Applied Optics, 19:4130.*

ZnS

Debenham, M. (1984). Refractive index of zinc sulfide in the 0.405–13 micron range, *Applied Optics*, 23: 2239.

Diamond

Edwards, D. F. and Ochoa, E. (1981). Infrared refractive index of diamond, *JOSA, 71*:607.

Field, J. E., ed. (1992). *The Properties of Natural and Synthetic Diamond*, Academic Press, New York.

Ge

Edwin, R. P., Dudermel, M. T., and Lamare, M. (1978). Refractive index of a germanium sample, *Applied Optics, 17*:1066.

Icenogle, H. W., Platt, B,. and Wolf, W. L. (1976). Refractive index and temperature coefficients of germanium and silicon, *Applied Optics, 15*:2348.

Ti Chalcogenide

Chalcogenide Glass Data Sheet #1173 (1969). Texas Instruments.

Strontium titanate

Rosch, S. (1965). The optical properties of fabulite, *Opt. Acta, 12*:253.

ZnSe

IR Materials, (1988). CVD booklet.

Dodge, M. J., and Malitson, I. H. (1975). Refractive index and temperature coefficient of index of CVD zinc selenide, NBS Special Publication #435:170.

Dodge, M. J. (1977). Refractive properties of CVD zinc sulfide, NBS Special Publication #509:83.

General References

Feldman, A., Horowitz, D., and Waslex, R. M. (1077). Refractive properties of infrared window materials, NBS Special Publication #509:74.

Ballard, S., McCarthy, K. and Wolf, W. (1959). Optical materials for IR instrumentation, Report #2389-11-S and suppl. of April 1961, University of Michigan, Willow Run Lab, Ann Arbor, MI.

Klocek, P., ed. (1991). *Handbook of Infrared Optical Materials*, Marcel Dekker, New York.

Herzberger, M. (1959). Color correction in optical systems and a new dispersion formula, *Optic Acta*, 6:197.

Morrissey, B. and Powell, C. (1973). Interpolation of refractive index data, *Applied Optics*, 12:1588.
Musikant, S. (1985). *Optical Materials*, Marcel Dekker, New York.
Palik, E. D. (1985). *Handbook of Optical Constants of Solids*, Academic Press, New York.
Sutton, L. E. and Stavroudis, O. N. (1961). Fitting refractive index data by least squares, *JOSA 51*:901.
Tatian, B. (1984). Fitting refractive index data with the Sellmeier dispersion formula, *Applied Optics*, 23:4477.
Tropf, W. J. (1995). Temperature dependent refractive index models for BaF_2 etc., *Optical Engineering*, 34:1369.
Wolf, W. L. (1978). Properties of Optical Materials, Sec. 7, *Handbook of Optics* (Driscoll, W. G., ed.) Optical Society of America, Mc-Graw-Hill, New York.
Wolf, W. and Zissis, G. eds. (1985). *The Infrared Handbook*, Environmental Research Institute of Michigan, Ann Arbor, MI.

INDEX OF REFRACTION PROGRAM (from file IRGLAS)

	7.50000	8.50000	10.00000	V	P
CAF2	1.36000	1.33878	2.39964	5.612	0.352
BAF2	1.43093	1.42032	1.40133	14.199	0.359
Strontium TIT	1.89166	1.76250	1.52036	2.054	0.348
LIF	1.24039	1.19206	1.10148	1.383	0.348
IRTRN2	2.22600	2.21619	2.19855	44.316	0.357
IRTTRN3	1.35999	1.33884	1.30022	5.669	0.354
IRTRN4	2.42036	2.41560	2.40737	108.931	0.367
CAD. Telluride	2.68559	2.68331	2.67965	283.520	0.383
Gallium Arsenide	3.28994	3.28552	3.27813	193.450	0.374
AMTIR-1	2.50491	2.50216	2.49758	204.803	0.375
AMTIR-3	2.61021	2.60726	2.60220	200.834	0.368
Ti Chalogenide	2.60856	2.60537	2.60008	189.333	0.376
Diamond	2.37597	2.37582	2.37564	4134.732	0.467
Silicon	3.41889	3.41832	3.41777	2161.307	0.505
Germanium	4.00641	4.00487	4.00339	995.663	0.511
ZnSe	2.41963	2.41485	2.40653	108.015	0.365
ZnS	2.22759	2.21784	2.20025	44.544	0.356
As2S3	2.39658	2.39123	2.38153	92.422	0.355

TEST GLASS FITTING

Test glass fitting of a system to be manufactured is described on pp. 17 and 18. This program automates this process. It assumes that a file containing the lens prescription is in format: Radius, Thickness, Material (see the DRAWING program for a description of this file format, p. 446). There also should be a file, (Hjglas, Coastal, etc.) that contains the listing of test glasses in ascending order of radii (with format F10.5.). the user is asked for the name of this file.

You are next prompted for the name of the file containing the prescription to be test glass fitted. Since all test glass data in the included company files is in inches, the user is then asked if his prescription is in mm. (If in mm, the test glass list is scaled to mm.) The user is then prompted for the radius tolerance. The radius tolerance is (Prescription radius – test glass radius)/radius.

All surfaces with radii falling within the above tolerance are changed to the test glass value. A new prescription file is created as L100.

A lens file Petzval is included on this disk. It is the Petzval lens shown in Fig. 5-2. Following is a listing of Petzval as well as its fitting to the Johnson Optical Labs test glass list (file Hjglass). A radius tolerance of 0.003 was used. The table indicates the seven surfaces that were fitted.

PROGRAM DESIGN A TRIPLET

This program computes the powers, surface curvatures, and separations for a triplet consisting of thin lenses. Four conditions are first solved for:

1. The lens assembly will have the desired effective focal length.
2. The Petzval sum will be zero.
3. Longitudinal chromatic aberration will be zero.
4. For a stop at the middle lens, the transverse chromatic aberration will be zero.

Petzval Lens f/1.4 14 Deg fov

	Radius	Thickness	Material	Diameter
0	0.000000	0.100000E+11		
1	1.42600	0.392050	LAK-N12	1.560
2	−2.58528	0.049378		1.560
3	−2.03990	0.142185	SF-8	1.480
4	3.10523	1.07933		1.280
5	1.05635	0.614096	LAK-N12	1.240
6	−2.09370	0.146425		1.240
7	−0.937755	0.111567	SF-54	0.800
8	7.11452	0.402289		0.800
1		1.4260		
2		2.5850		
3		2.0346		
4		3.1091		
6		2.0930		
7		0.9350		

7 surfaces have been test glass fitted.

To simplify the computation, it is assumed that the two outer lenses consist of the same material. Refer to pp. 55–57.

The user is asked to input $N(A)$, $V(A)$, $N(B)$, $V(B)$, F, TR, where

$N(A)$, $V(A)$ are the refractive index and V value for the first and last thin lens.

$N(B)$, $V(B)$ are the refractive index and V value for the center thin lens.

F is the desired effective focal length.

TR is the ratio of the two air spaces. $TR = T2/T1$.

Output is the power of the thin lenses, $T1$, $T2$, and the BFL. With these values fixed, the surface curvatures are found with a least squares minimization routine. Output are surface curvatures and the sum of the squares of the three third-order aberrations that were minimized; spherical, coma, and astigmatism.

Program Optics   463

The following example uses SK16 and LLF1 glasses with TR=1.

Thin Lens Triplet

1.62286	60.08000	1.55099		45.47000	100.0000	1.00000
PHA	PHB	PHC		T1	T2	BFL
0.03347	–0.06525	0.03480		3.79942	3.79942	96.20059
CV1	CV2	CV3	CV4	CV5	CV6	SUM
0.0117640	–0.0419798	–0.1102762	0.0081439	–0.0585666	–0.1144330	0.00000

PROGRAM ZOOM CAM

This program computes the cam data to make the cams for a mechanically compensated zoom lens. The following input data is requested from a file with the following order;

    First line   Heading
    Next M lines   Values of $T1$ and $T2$ in format, 2E11.5.
                         Data may be in fixed format or separate data with commas as in file Lenscam

You are then asked

> The starting focal length or magnification FS. This is the starting effective focal length (or magnification). The last value computed is $Z*FS$.
> The zoom ratio Z.
> The scale factor; the cam data is to be multiplied by YS.
> YS is the scale factor applied to the cam data. Input 1 for a full size cam.
> A large number (say 10) is useful to make a very large master cam and then reduce the production units down by pantograph reproduction of this master cam.
> Sum of the zoom spaces = $T1 + T2 + T3$, where (5.21591 in this example)

T1 is the space between the A and B groups
T2 is the space between the B and C groups
T3 is the space between the C and D groups

For a zoom lens with four zoom spacings, input −1 for the sum of zoom spaces.

Angle. This is the angle through which the cam is to be turned. It is the total angular rotation of the cam in degrees. it is divided into $N(M-1)$ equal increments.

The number of coordinates N. Each input zoom position is then divided into N logarithmically equally spaced positions.

Thus the number of coordinates calculated is $N(M-1) + 1$.

By a least squares fitting, the input data (M items of $T1$ and $T2$) is fitted to polynomials of the form

$$Y = A + BX + CX^2 + DX^3 + EX^4 + FX^5$$

The coefficients A, B, C, D, E, and F are computed and printed. Next a table of the input values vs. the computed values is printed followed by the sum of the absolute errors. A table of $T1$, $T2$, and $T3$ is then printed. Due to the large amount of data created, this data is not displayed but is instead always printed. The printer then must be on. Values of Y are computed for $X=1$ to $X=M$.

Note: These values of $T1$, $T2$, and $T3$ differ from values computed using the Zoomsp program. Zoomsp values are first-order computations, whereas these values are interpolated from data taken from the optimization program. Since the program does not assign any physical significance to the spacing values $T1$, $T2$, and $t3$, any two of these values may be input.

Cam for In-Water Zoom Lens (file Lenscam)

0.47901,4.39650
0.63001,3.19092
0.72867,2.16562
0.75237,1.32624
0.67408,0.686810
0.46592,0.267190
0.10615,0.086140

Cam for In-Water Zoom Lens

Number of Zoom Positions	7
Number of Points	5
Range of Zoom	4.00000
EFL/MAG	0.63015
Y Scale	1.00000
Sum	5.21591
Angle	270.00000

Zoom Coordinates

0.47901	0.63001	0.72867	0.75237	0.67408	0.46592	0.10615
4.39650	3.19092	2.16562	1.32624	0.68681	0.26719	0.08614

Coefficients of Polynomial 1 in ascending order of powers of X are
    0.28738E+00   0.21165E+00   −0.21373E−01   0.21807E−01
    −0.87561E−03   0.43958E−04

Coefficients of Polynomial 2 in ascending order of powers of X are
    0.57914E+01   −0.14970E+01   0.10754E+00   −0.63969E−02
    0.98409E−03   −0.39542E−04

T1		T2	
0.47901	0.47901	4.39650	4.39650
0.63001	0.3003	3.19092	3.19089
0.72867	0.72861	2.16562	2.16569
0.75237	0.75244	1.32624	1.32615
0.67408	0.67402	0.68681	0.68688
0.46592	0.46594	0.26719	0.26716
0.10615	0.10615	0.08614	0.08614
ERROR	0.00024	0.00030	

Cam for In-Water Zoom Lens

EFL/MAG	T1	T2	T3	Angle
0.6301	0.4790	4.3965	0.3404	0.00
0.6600	0.5126	4.1408	0.5625	9.00
0.6912	0.5447	3.8924	0.7788	18.00
0.7239	0.5750	3.6514	0.9896	27.00
0.7581	0.6035	3.4175	1.1949	36.00
0.7939	0.6300	3.1909	1.3050	45.00
0.8315	0.6545	2.9714	1.5899	54.00
0.8708	0.6768	2.7591	1.7799	63.00
0.9120	0.6967	2.5540	1.9651	72.00
0.9551	0.7141	2.3562	2.1456	81.00
1.0003	0.7286	2.1657	2.3216	90.00
1.0476	0.7402	1.9826	2.4931	99.00
1.0972	0.7486	1.8069	2.6604	108.00
1.1490	0.7536	1.6389	2.8234	117.00
1.2034	0.7549	1.4786	2.9824	126.00
1.2603	0.7524	1.3261	3.1373	135.00
1.3199	0.7459	1.1817	3.2884	144.00
1.3823	0.7349	1.0454	3.4356	153.00
1.4477	0.7195	0.9174	3.5791	162.00
1.5162	0.6992	0.7978	3.7188	171.00
1.5879	0.6740	0.6869	3.8550	180.00
1.6630	0.6436	0.5847	3.9876	189.00
1.7416	0.6078	0.4915	4.1166	198.00
1.8240	0.5663	0.4074	4.2422	207.00
1.9103	0.5191	0.3326	4.3642	216.00
2.0006	0.4659	0.2672	4.4828	225.00
2.0952	0.4067	0.2113	4.5979	234.00
2.1943	0.3412	0.1651	4.7096	243.00
2.2981	0.2693	0.1288	4.8178	252.00
2.4068	0.1910	0.1025	4.9225	261.00
2.5206	0.1061	0.0861	5.0236	270.00

PROGRAM ZOOM SPACINGS

This program calculates the zoom spacings for either an optically compensated or a mechanically compensated zoom lens. The user is first asked for the file name containing the lens prescription. (The program reads lens prescription data per format 6X,4G12.6.)

> First line   Heading
> Next lines   lens values of Radius, Thickness, N, V in format 6X,4E12.5. (Note that V, the Abbe V value, is not used by the program and so is an optional input value.) Data may be in fixed format as in the example, or separate data with commas.

You are next asked if this is an optically or mechanically compensated system. The variable air spaces referred to assume that the object space is $T0$, the first lens thickness is $T1$, etc. These surface numbers are integers; do not use a decimal point!

> For a mechanically compensated system, you are requested to enter
> The surface number of the first surface of the B group
> The surface number of the last of the B group
> The surface number of the last of the C group
> (Thus there are three variable air spaces.)
> For an optically compensated system, you are first asked for the number of variable air spaces, then the actual air space numbers.

You are asked for the zoom ratio referenced to the input system (enter a number with a decimal point) and the number of zoom steps (integer input.)

> For mechanically compensated systems, output is the zoom spacings vs. focal length or magnification.
> For optically compensated systems, output is lens movement and image errors vs. focal length or magnification. In the printout, the last column labeled TH is the paraxial BFL.

All computations are first order.

The example is contained in file Lensz and is similar to that shown in Fig. 35-1.

Values for 10 steps are displayed. In each step, the focal length (or magnification) is multiplied by the previous step; in this case, the multiplier is $\sqrt[9]{10} = 1.29155$.

Lens Group Surfaces	A 1–5	B 6–10	C 11–15	D 16–24

A plot of the lens movements is shown as Fig. A3.

10 × ZOOM LENS   (from file Lensz)

	Radius	Thickness	Index	V
0	0.000000	0.100000E+10	1.00000	0.000000
1	7.98401	1.27064	1.69401	54.4531
2	−6.65062	0.236048	1.67764	31.9558
3	13.8143	0.248805E-01	1.00000	0.000000
4	4.35143	0.574170	1.62689	46.6795
5	10.3710	0.195187E-01	1.00000	0.000000
6	8.12216	0.141638	1.69401	54.4531
7	1.22185	0.531738	1.00000	0.000000
8	−3.75639	0.157316	1.69401	54.4531
9	1.76584	0.333420	1.67764	31.9558
10	0.000000	4.54061	1.00000	0.000000
11	2.79956	0.269167	1.68083	54.8953
12	−9.86742	0.409600E-01	1.00000	0.000000
13	1.63383	0.117940	1.72311	29.2765
14	0.889158	0.510012	1.68083	54.8953
15	12.4418	0.830029E-01	1.00000	0.000000
16	0.000000	0.454536E-01	1.00000	0.000000
17	−1.90622	0.119059	1.72055	47.6568
18	−1.53582	0.984289E-01	1.72311	29.2765
19	1.11258	0.104213	1.00000	0.000000
20	−2.74528	0.256995	1.69416	30.9285
21	−1.37624	0.296643E-01	1.00000	0.000000
22	1.32400	0.277150	1.57487	57.2348
23	−1.85516	0.154994	1.72311	29.2765
24	−2.01437	1.28198	1.00000	0.000000

Program Optics

Focal Length	T5	T10	T15
0.59132	0.01952	4.54061	0.08300
0.76372	0.65889	3.84316	0.14109
0.98638	1.22079	3.21548	0.20686
1.27396	1.71451	2.64728	0.28134
1.64539	2.14820	2.12925	0.36568
2.12510	2.52904	1.65290	0.46119
2.74467	2.86332	1.21043	0.56939
3.54488	3.15652	0.79456	0.69205
4.57838	3.41329	0.39846	0.83138
5.91321	3.63665	0.01549	0.99100

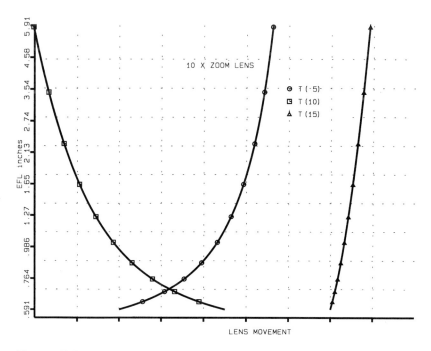

Figure E-3 Lens movement.

Index

Achromat
  doublet, 45
  as an extender, 94–95
  pseudo, 53
  reduced secondary color, 50
  thin lens, third order, 45–56, 440
Achromatic prism (*see* Achromatic wedge)
Achromatic wedge 287–289, 434
Aerial photography lenses, 313-322
Afocal optical systems, 157–168, 218–219, 346–350, 355–359
Airy disc, 226
Anamorphic systems, cylinder, 297–301
Anamorphic systems, prism, 301–303
Antireflection films, 11–13
Apochromat microscope
  objective 137
  telephoto, 92–93
  telescope objective, 52

Aspheric surfaces, 6–8, 189, 193, 197, 443

Baffles, 185
Barlow lens (negative achromat), 50, 92
Best form lens, 212, 442
Binocular, Galilean, 160, 162
Bi-ocular lens, 283–284
Bounds, 17

Cassegrain mirror system, 181–183, 224–225, 350–355, 449
Catadioptric lenses, 139–140, 181–196, 224–225
Cemented surfaces, 11, 249
Cinegon lens, 99–100
CinemaScope, 297
Computer programs for lens design
  commercially available, 437–438
  use, 30

471

Index

Condenser, 305-312, 442
Conic section lenses, 273–274
Copy lens, 172-174, 401-407

Designs, list, 9–12
Diopter, prism, 287
Door viewer, 167
Depth of field, 37-38
Diffraction limit, 36-37
Distortion, ix
Document scanner, 233–235
Drawings, lens, 27–29,
   445–448

Elements, changing number of,
   15
Ellipsoidal mirror, 311–312
Edge and center thickness,
   17
Endoscope, 241-244
Enlarging lenses, 245-248
Erfle eyepiece, 127
Eyepiece relay, 177–178
Eyepieces, 121–132
   magnification, 123
Eye pupil diameter, 121

Field angles, 5
Film formats, 429–430
Filters, aerial photography,
   313
Fish-eye lenses, (see Wide
   angle lenses)
Flange distances, 433
FLIR (forward looking
   infrared), 213–215
Fluorite objective, 137

Focusing
   eyepiece, 122
   periscope, 206
   zoom, 332, 375
F theta (see Scan lenses)
Fringes (accuracy/irregularity,
   28

Glass variation, 9–11
Gauss, double, 75–83
Gradient index lenses, 419–428

Heads up display (HUD),
   279–285
   specifications, 279
Heliar, 65–66
Hyperhemisphere bound, 113

Illumination systems, 305–312
Index of refraction
   calculations, 8, 453
   immersion oil, 142
   polynomial fit, 453
   Schott equation, 9, 145, 454
   Sellmeier formula, 9, 454
   UV transmitting glasses,
      222
   water, 147
Infrared lenses, 211–220
   achromat doublet, 46
   Cassegrain, 183–184
   triplet, 59–61
   zoom, 350–355
Inverted telephoto, 97–103, 195
Irtran, 458

Kellner eyepiece, 124

Laser beam expander, 157–161
  focusing lenses, 271–277
  lines, 41
  optics, 39
  scan lens, 235–236
Least squares, optimization, 4
Lister-type objective, 134
Lithography, optical 227–230

Mechanical properties, 411–412
Melt data fitting, 19–20
Merit function, 2
Microfiche, 327
Microscope objectives, 133–143
  oil immersion, 141–142
  reflecting, 139–140, 195–196
  zoom, 346–350, 360–370
MTF, 450
  wavelength vs. weights, 8

Narcissus, 33–34
Number of elements, changing, 15
Numerical aperture, 170

Optics program, 439–470
Optimization methods, 3
Orthonormal, optimization, 4

Panoramic camera, 118–119
Parameters, variable, 16–17
Pellicle, 291
Periscope lenses, 203–209
Petzval lens, 71–74

Photographic (*see also* Aerial photography lenses)
  lenses, 31–32
  Cassegrain, 187–188
  zoom, 375, 351
Plastics
  4 × Galilean, 162–163
  projection lens, 258–259
  video disk, 274
Plossl eyepiece, 127
Port
  concentric dome, 151–152
  flat plate, 147-150
Power changer, 163-164
Prism, achromatic (*see* Achromatic wedge)
Profile projection, 266
Projection lenses, 249–263, 452
  displaced, 253
  LCD, 260–262
  microprojection lenses, 327–329
  requirements, high power, 249
  slide, 68–69
  variable focal length, 409
  zoom, 396
Pupil aberration, 105
Pupil shift, 13–14

Radiation resistant lenses, 323-326
Ray pattern, 4–5
Reflector, xenon arc, 311–312
Refractive index (*see* Index of refraction)
Relay systems, 169–189

Retrofocus lens (*see* Inverted telephoto)
Rifle sight, 175–176, 335-360
Ritchey–Chretien, 197–200
Rod lenses, 242
Rotary prism camera, 294–295

Scan lenses (F theta), 233–239
Schmidt camera, 193–194
Secondary color, 34–36
Sparrows criterion, 37
Spherical aberration, minimum, 212, 442

T number, 204
Telecentric lenses, 265–270
Telephoto lens, 85–95
   first order solution, 85
Telescope objective, 47–53, 189-190
Tessar, 67–68
Test glass fitting, 17–19, 462
Thermal problems, 20–21
Thermal properties, 435
Tilted plate, 294
Tolerances, 21–26
Transmittance
   in atmosphere, 211
   in ocean water, 146
Triplet lens (air spaced), 55
   modifications, 65–70
   radiation resistant, 323-325
   third order solution, 55–57, 463

TV-type lenses, 32
   zoom, 341, 371, 380
Ultrafiche, 327
UV lenses, 34, 221–231
   achromat doublet, 46
   reflecting objective, 139–140

Variable focal length lenses, 409–417
Video disk lens, 271–272
Viewfinder, Albada, 165
   eyepiece relay, camera viewfinder, 177–178
Vignetting, 13–14

In-water lenses, 145–156
Wedge plate, 291–293
Wide angle lenses (*see* Inverted telephoto)
   very wide angle, 105-120
Whitworth thread, 133

Zoom lenses
   cams, 464–467
   first order theory, mechanical, 331–336
   first order theory, optical 337–340
   Mechanically compensated, 387–400
   optically compensated, 387–400
   spacings, 468–470